AF412729

Diffraction X-ray Optics

Diffraction X-ray Optics

A I Erko

Russian Academy of Sciences,
Chernogolovka, Moscow region

V V Aristov

Russian Academy of Sciences,
Chernogolovka, Moscow region

B Vidal

Université d'Aix-Marseille III

Institute of Physics Publishing
Bristol and Philadelphia

British Library Cataloguing-in-Publication Data

A catalogue record for this book is available from the British Library.

ISBN 0 7503 0359 X hardback

Library of Congress Cataloging-in-Publication Data
Erko, A. I. (Alekseĭ Ivanovich)
    Diffraction X-ray optics / A.I. Erko, V.V. Aristov, B. Vidal ;
[translated by S. Chomet].
        p.   cm.
    Includes bibliographical references and index.
    ISBN 0-7503-0359-X (alk. paper)
    1. X-rays--Diffraction. 2. X-rays--Diffraction--Industrial
applications. I. Aristov, V. V. (Vitaliĭ Vasil'evich) II. Vidal,
Bernard. III. Title.
QC482.D5E75   1996
621.36'73--dc20   96-11499
                                                                 CIP

Translated by S Chomet
Translation editor: Professor Michael Hart, University of Manchester

Published by Institute of Physics Publishing, wholly owned by The Institute of Physics, London

Institute of Physics Publishing, Techno House, Redcliffe Way, Bristol BS1 6NX, UK

US Editorial Office: Institute of Physics Publishing, The Public Ledger Building, Suite 1035, 150 South Independence Mall West, Philadelphia, PA 19106, USA

Typeset in TeX using the IOP Bookmaker Macros
Printed in the UK by Bookcraft Ltd, Bath

Contents

Preface

This book is based on the theoretical, experimental and technological experience obtained in the Institute of Microelectronics Technology of the Russian Academy of Sciences (IMT RAS) and the Laboratory of Electromagnetic Optics (LOE) of Marseilles University in the field of high-resolution X-ray optics fabrication. Since 1983, investigations carried out here have led to the creation of the well established technology and diagnostics for X-ray optics fabrication including multilayer and crystal Bragg–Fresnel optics. The wide spectrum of X-ray optical elements produced in IMT RAS and LOE, such as zone plates, free-standing transmission diffraction gratings, multilayer mirrors and supermirrors, Fourier masks, Bragg–Fresnel lenses and multilayer diffraction gratings, have been used in experimental investigations in several laboratories in Russia, Europe and Japan. Thus, the first experimental tests of Fourier imaging techniques in X-rays were done at the Photon Factory (Tsukuba, Japan). The elliptical Bragg–Fresnel lens was tested first at the Siberia-1: the synchrotron source at the Kurchatov Institute (Moscow, Russia). Experiences collected between 1983 and 1989 were summarized in the book of V V Aristov and A I Erko *Roentgen Optics*, published in 1990 by Nauka (in Russian). However, since that time a lot of new theoretical and experimental data have been obtained: several other laboratories were involved in the production, tests and application of Bragg–Fresnel optics. The term Bragg–Fresnel optics, introduced by V V Aristov in 1987, is now commonly used in scientific literature and as a section title at international conferences. A fruitful international co-operation between IMT RAS and the French 'Centre National de la Recherche Scientifique' (CNRS) laboratories has led to the development of an advanced technology and theory of multilayer gratings and Bragg–Fresnel lenses. Since October 1991 the first Bragg–Fresnel fluorescence microprobe with submicron resolution has been in operation at LURE, Orsay, France. Several experiments have been performed using this microprobe at the European Synchrotron Radiation Facility (ESRF) in Grenoble. The future of Bragg–Fresnel optics seems to be very promising in synchrotron beam monitoring, plasma diagnostics, microdiffraction etc.

Proceedings of the Fourth International Conference on X-ray Microscopy published in 1994 (edited by V V Aristov and A I Erko) collects some new results in the field of X-ray, as well as Bragg–Fresnel optics. But publication of a new book with a detailed description of the basic principles of X-ray diffraction optics with recent results and applications seems to be required.

The authors are very grateful to our colleagues from IMT RAS and CNRS France for their great contribution to this work.

A I Erko, V V Aristov and B Vidal

Introduction

The use of nanometre radiation as a tool for microscopy and nanometre technology is developing very rapidly. This can be attributed to the development of synchrotron radiation sources and the progress in optical elements and methods for X-ray beam control such as focusing and modulation. The future of this technique is also very promising with the advent of X-ray lasers.

X-ray optic systems based on total external reflection of radiation from a metal surface are now widely available. However, the necessity of using grazing incidence angles of radiation substantially limits the performance of such optics. The development of microstructurization methods, namely, the fabrication of planar submicron structures with sizes of X-ray wave order, deposition and growth of thin films of different materials, enabled the fabrication of diffraction optical elements of the nanometre range; these are mainly Fresnel optics elements. The development of multilayer interference mirrors for the nanometre range, providing efficient reflection at angles close to normal incidence, was a great step forward. And, finally, the present-day stage of developing new optics for X-ray beams calls for the creation of efficient focusing elements with the structure of three-dimensional Fresnel zones: Bragg–Fresnel optics.

In 1963, Yu N Denisiuk published the idea of using a combination of volume Bragg diffraction and Fresnel or Fraunhofer diffraction for the recording and reconstruction of optical holographic images. The first proposal for the application of this technique to X-rays was made in 1982 by V V Aristov *et al*. The idea for getting X-ray focusing by combining Bragg reflection and diffraction by a Fresnel zone is a recent one. In 1986 such a principle was applied to the fabrication of artificial structure of volume holograms on multilayer and crystalline substrates. The Bragg–Fresnel lens (BFL) keeps the advantage of the Fresnel zone plates, which offers a high spatial resolution when used in transmission. Compared to transmission free standing zone plates, BFL has a high mechanical stability demonstrated on multilayer X-ray mirrors or crystals. Such a high mechanical stability becomes especially important with the advance of ultra-bright sources.

These structures are the first elements of microphotonics as they allow for transforming the information of electrical, optical or acoustic signals into X-ray beam modulation. Success in the development of Bragg–Fresnel optics points to the formation of a new trend connected with the use of nanometre radiation in diagnostics, transmission and processing of information, namely, X-ray electronics as a branch of the general science of electromagnetic radiation control, i.e. photonics. This trend is based on the development of powerful radiation sources in the nanometre range of synchrotrons and storage rings, X-ray lasers in future, and on the methods of microelectronics technology that enable the fabrication of structures of X-ray devices with submicron and subnanometre element sizes.

The purpose of this book is to give a systematic description of the common mathematical background properties and performances of diffraction X-ray optical elements based on artificially made microstructures: plane zone plates, two- and three-dimensional gratings, multilayer mirrors and three-dimensional Fresnel structures. Kinematic diffraction theory and classical optics theory are used as the basis for this description. In this book we publish not only theoretical models and estimations. Most of the calculated data are tested experimentally; we give a detailed description of instrumentation and experimental conditions. It should be stated that most of the theoretical results obtained for multilayer structures in the kinematic approximation can be used only as a qualitative estimation for crystal optics. A more precise dynamical theory must be applied to the description of a crystal structure as well as crystal Bragg–Fresnel optics.

Finally, this is the first book with a detailed description of the main principles, technology and some applications of the Bragg–Fresnel multilayer optics as well as X-ray Fourier optics. The crystal BFL is discussed in more detail in the works of Aristov and Snigirev.

The book consists of six chapters. Current information on the optical properties of materials in the X-ray wavelength range and the methods of measurement of the optical constants are considered in Chapter 1. Optical devices based on the effect of total external X-ray reflection from polished surfaces are only outlined. This is due to the fact that several comprehensive monographs have already been published on the subject. Much attention is paid to the analysis of characteristics of the Fresnel zone optics (Chapter 2) and the methods of X-ray coherent transmission (Chapter 3). Resolution, efficiency and aberration properties of Fresnel and Fourier optics are analysed. Practical examples of applications of this kind of optics are illustrated. With the consistent consideration of Bragg–Fresnel optical elements on the basis of multilayer mirrors, X-ray optics of a new type are first described in Chapter 4.

Development of the microstructurization methods, namely, fabrication of planar submicron structures with sizes of X-ray wave order, deposition and growth of thin films of different materials, has enabled fabrication of diffraction optical elements of the nanometre range. The book is distinguished by detailed analysis of technological aspects of the fabrication of such structures. This problem is considered in Chapter 5. Classical and new methods in multilayer mirror technology are described.

The last chapter is devoted to an application of diffraction focusing optics in microscopy and microbeam instrumentation. Physical principles of the main microscopy methods using X-ray photons are outlined. As a practical example of the real microprobe a detailed description of the LURE-IMT fluorescence microprobe is given.

Notation

δ	refractivity
f_1, f_2	atomic scattering factors
μ	ionization cross section
W	energy
Z	atomic number
N_a	Avogadro number
N_g	number of atoms of type g per unit volume
N_{f_g}	mean atomic scattering factor
n	complex refractive index
λ	wavelength
r_c	classical electron radii
$\hbar$	Planck's constant
β	absorption index
χ	'phase' factor
θ_c	critical angle
$\hat{R}$	intensity reflection coefficient
μ_t	linear absorption coefficient
θ	grazing angle
S	complex reflection coefficient
$\phi(W_c)$	phase shift
t	thickness of the material
δ_{x_i}	spatial resolution of Fraunhofer hologram
δ_r	registrator resolution
d_s	source size
ρ	material density
$J_n(x)$	Bessel function
$I(x)$	intensity
r_n	Fresnel radii
m	diffraction order
E	wave amplitude
ΔL	optical path difference
F	focal distance
t_π	thickness of the π phase-shift

t_{opt}	optimal zone plate thickness
k	rigidity parameter
Λ_0	period of modulation
d	structure period
ℓ	number of layers
$\hat{h}$	amplitude of oscillations
ξ, η	elliptical coordinates
λ_c	de Broglie wavelength
C_s	spherical aberration coefficient
d_p	electron probe size
η_2	back-scattering coefficient
τ_0	exposure time
ϕ_{lin}	linear image spread function
t_{exp}	extinction depth
$\Phi(\lambda)$	photon flux

1

Absorption and refraction of nanometre waves

In this chapter, we present the physical principles underlying the interaction of X-rays with matter in the X-ray wavelength range 0.1–20 nm (10 keV–50 eV). We use atomic scattering factors to characterize the optical properties of materials and then examine the connection between these factors and the refractive indices and absorption coefficients that are commonly used in optics. Materials suitable for high-efficiency X-ray optics are then chosen and classified on the basis of a survey of published data. The chapter ends with a review of experimental methods for the determination of the optical constants of materials in the X-ray range.

1.1 FACTORS GOVERNING THE INTERACTION OF X-RAYS WITH MATTER

The main emission (resonance) frequencies of bound electrons in most elements of the Periodic Table correspond to wavelengths in the range 0.01–10 nm. In contrast to the optical range in which the refractive index is always greater than unity, the X-ray refractive index differs from unity by a small negative amount δ. In the region of a fundamental absorption edge of an atom, the refractivity δ changes sign in a very narrow wavelength range, and this is accompanied by a sharp change in the absorption coefficient [1].

The physics of the interaction between hard X-rays (wavelength $\lambda < 0.1$ nm) and matter is described in some detail in James' book [2]. If we follow the description of the interaction between an atom and an X-ray photon given in [3], we can describe the interaction in the case of soft X-ray photons in terms of coherent scattering and photoelectric absorption. Incoherent scattering may be neglected. Thus, the scattering of an X-ray photon by an atom may be represented by the complex scattering amplitude

$$E = E_0(f_1 + \mathrm{i}f_2)(r_\mathrm{e}/R)P(2\theta) \tag{1.1}$$

where $f_1 + \mathrm{i}f_2$ is the complex scattering factor, $r_\mathrm{e} = e^2/mc^2$ is the classical radius of the electron, R is the distance to the point of observation, $P(2\theta)$ is

"

the polarization factor, E_0 is the amplitude of the incident field, 2θ is the angle between the direction of incidence and the direction of scattering and f_1, f_2 are factors describing coherent scattering (refraction) and photoelectric absorption, respectively.

For radiation of wavelength greater than the dimensions of an atomic shell, i.e. for soft X-rays, it may be considered that all the electrons in a shell scatter in phase and (this is very important) f_1 and f_2 are independent of angle. The only parameter that remains angle dependent in (1.1) is therefore $P(2\theta)$. It is equal to unity when the polarization vector of the incident radiation is perpendicular to the plane of scattering, and to $\cos 2\theta$ when the polarization vector lies in this plane. For unpolarized radiation, the scattered intensity is given by

$$I(\lambda, \theta) = I_0(r_e/R)^2(f_1^2 + f_2^2)[1 + \cos^2(2\theta)]/2. \tag{1.2}$$

Calculations based on the relativistic quantum theory of dispersion [4] have shown that f_2 is directly proportional to the photoionization cross section μ_a and that f_1 can be expressed in terms of the ionization cross section as follows:

$$f_2(W) = \frac{W\mu_a(W)}{2\pi r_e \hbar c} \tag{1.3a}$$

$$f_1(W) = Z + \frac{1}{\pi r_e \hbar c} \int_0^\infty \frac{W_c \mu_a(W_c) \mathrm{d}W_c}{W^2 - W_c^2} + \Delta f_2 \tag{1.3b}$$

where Δf_2 is a relativistic coefficient that is very small for energies of the order of 1 keV, Z is the atomic number of the element, $\hbar$ is Planck's constant divided by 2π and W_c is the energy of a photon. The values of f_1 and f_2 in the range between 100 eV and 2 keV were calculated in [5] from experimental data on the photoionization cross section μ_a.

For systems consisting of different atoms, the mean atomic scattering factor for soft X-rays is given by

$$\overline{Nf_1} = \sum_g N_g f_{1g}, \qquad \overline{Nf_2} = \sum_g N_g f_{2g} \tag{1.4}$$

where $\overline{Nf_{1,2}}$ is the mean atomic scattering factor per unit volume and N_g is the number of atoms of type g per unit volume.

It is therefore clear that, in the soft X-ray range, the scattering factors can be calculated from the photoionization cross sections. At photon energies above 50 eV they are independent of the scattering angle and of the state of the atomic system (if we exclude the absorption edges).

The complex refractive index is given by

$$\hat{n} = 1 - \delta + \mathrm{i}\beta \tag{1.5}$$

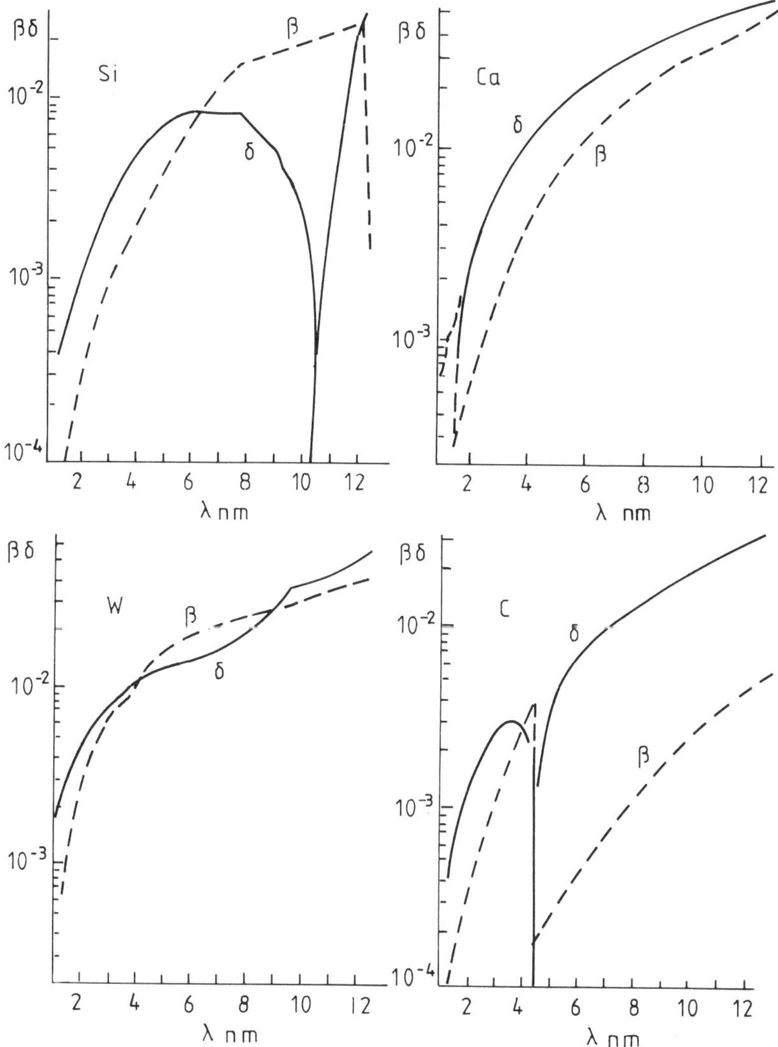

Figure 1.1 Optical constants of silicon, cobalt, tungsten and carbon, calculated from the data reported in [5].

where $\delta = r_e\lambda^2 N_g f_1/(2\pi)$ and $\beta = r_e\lambda^2 N_g f_2/(2\pi)$.

The real part of the refractive index is difficult to measure. Lukirskii *et al* [6] were the first to initiate such measurements, but there is still a lack of such data for most elements. Measurements performed at the shorter wavelengths in this range have shown that there was relatively good agreement between experimental and theoretical values of δ and β [7]. Measurements at longer wavelengths encounter considerable difficulties because practically all materials have high absorption coefficients at these wavelengths, so that classical methods cannot be employed. There have been several recent publications [8] suggesting that multilayer structures could be used in measurements of optical constants. However, this method is very sensitive to fabrication errors in the multilayer structures, and provides only a qualitative characterization of the behaviour of refractivity at different wavelengths.

It is thus clear that the calculated real part of the refractive index does not take into account the actual structure of matter, i.e. the interaction between atoms in the crystalline (or amorphous) matrix, and satisfactorily describes the optical properties of materials only well away from absorption edges. All this must be borne in mind when the values of f_1, f_2 are used to calculate the optical properties of any particular material.

Figure 1.1 shows the functions $\delta(\lambda)$ and $\beta(\lambda)$ for four different elements. These curves were calculated from (1.5) with f_1 and f_2 taken from the tables in [5]. These graphs show that, for both light and heavy elements, the real part of the complex refractive index differs from unity by less than 0.01 at wavelengths up to 20 nm. The absorption β is of the same order, showing that all elements exhibit strong absorption in the nanometre wavelength range. However, the real and imaginary parts of the refractive index are found to be very different near an absorption edge. Although the values of f_1 in such regions cannot be calculated very accurately from the values of μ_a, the results obtained suggest that there are transparency windows near absorption edges, in which the material can be regarded as a 'phase medium'. If we use the parameter $\chi = \delta/\beta$ as a measure of this 'transparency', we can determine the wavelength window λ_1 to λ_2 in which the material behaves as phase medium. Examples of such materials and the corresponding values of λ_1 and λ_2 (taken from [9]) are listed below.

Material	Au	Ag	Cu	Si	Al	C
λ_1 (nm)	0.2	0.46	0.44	0.35	1.45	4.8
λ_2 (nm)	0.23	0.88	0.54	2.0	2.25	8.6

Materials with reasonable 'phase properties' are mixtures of several elements, e.g. chalcogenide glasses, polymers *etc*. The refractive indices of such media are more difficult to calculate than the indices of the individual component elements. Nevertheless, the problem has to be solved because there is a whole series of materials of complex composition whose physical and chemical properties can be modified by exposure to electron, ion and X-ray beams, and this can be exploited in the fabrication of X-ray optical elements.

Table 1.1

Material	Thickness μm	λ nm
Silicon doped with boron	1–5	0.7–1.5
Beryllium	12	0.7–1.5
Mylar	2–25	<2.0
		4.4–6.0
Polyimide	12	0.7–1.5
Si_3N_4	0.1–0.2	<4.0
$Si_3N_4 + SiO_2$	0.3–0.4	<2.0
Al_2O_3	3–5	8–15
Polypropylene	1–25	<25
Polycarbonate	2	<2.0
		4.4–6.0
Polyvinyl	0.4	<3.0
		4.4–7.5
Aluminium	6	0.8–1.4
Nitrocellulose	1–5	<2.0
		4.4–6.0
Parylene	0.2–3	<7.0

Polyimide is a very interesting material for X-ray optical elements. A suitably treated film of this material exhibits considerable strength and thermal stability (up to 400 °C). At the same time, the polymer has a transparency window near the K-edge of carbon. This property of polyimide is widely exploited in the production of transparent membranes that constiute the basis of X-ray optical elements for the nanometre wavelength range. Table 1.1 lists the transparency windows for a number of materials used as transparent templates.

1.2 METHODS OF MEASURING OPTICAL CONSTANTS

We have already noted that the basic characteristic of the optical properties of materials is the photoionization cross section μ_a. The refractive index $1 - \delta$ and the absorption β can be calculated from (1.3b). Because the relevant absorption coefficients are high, direct methods for the determination of the photoionization cross section in the nanometre range that rely on transmission data are relatively complicated and require thin (thinner than 0.3 μm) films of known purity and thickness. It is important to remember that the optical properties of films can be very different from those of materials in bulk, and may depend on the method used to produce them (epitaxy, thermal deposition, etching *etc.*). The particular feature of the nanometre range is that it lies between the well studied ultraviolet and hard X-ray regions. This means that the optical constants of solids can be determined as equally well by methods developed for hard X-rays and purely optical methods [10]. The experimental possibilities have been vastly expanded in relation to measurements of the photoinization cross section since the advent

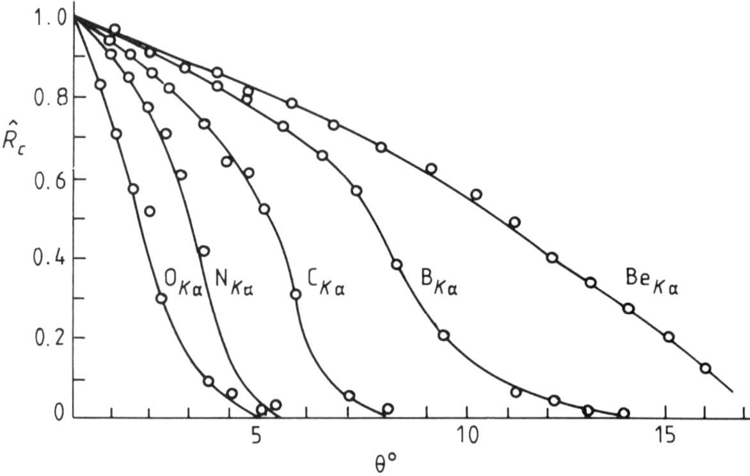

Figure 1.2 X-ray reflection coefficient of titanium as a function of the grazing angle for different wavelengths [14].

of synchrotron sources of radiation. Indeed, high-precision measurements of this kind have now been made on a large number of intermediate and heavy elements [11–13].

Measurements of the real part of the refractive index $1 - \delta$ present a particularly difficult problem. We shall now examine in detail some of the methods used for this purpose.

The pre-eminent method relies on the determination of the angular distribution of total external reflection. Assuming that absorption can be neglected and that grazing angles are employed, we find that the refractive index is given by the relatively simple formula

$$\hat{n} = 1 - \delta = \cos\theta_c \approx \sqrt{1 - \theta_c^2/2} \tag{1.6}$$

where θ_c is the critical angle, so that

$$\delta = \theta_c^2/2. \tag{1.7}$$

The critical angle for total external reflection corresponds to the maximum derivative of the reflection curves $\hat{R}_c(\theta)$. Figure 1.2 [14] shows the reflection curves for titanium for different wavelengths of characteristic radiation. The method described in [14, 15] can be used for more accurate measurements capable of producing absorption data. These two publications present systematic measurements of optical constants of different materials in the wavelength range 0.7–11.3 nm.

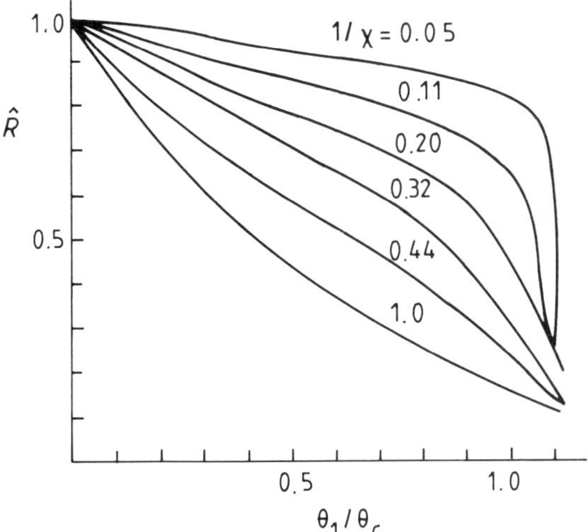

Figure 1.3 Calculated parametric curves used to determine optical constants [14].

The real and imaginary parts of the refractive index can be calculated from the Fresnel formulas

$$S_\| = \frac{\hat{n} \sin \theta_1 - \sin \theta_2}{\hat{n} \sin \theta_1 + \sin \theta_2} \qquad S_\perp = \frac{\sin \theta_1 - \hat{n} \sin \theta_2}{\sin \theta_1 + \hat{n} \sin \theta_2} \qquad (1.8)$$

where $S_\|$ and $S_\perp$ are the amplitude reflection coefficients for the incident wave polarized parallel and perpendicular to the plane of incidence, θ_1 is the angle of incidence, θ_2 is the angle of refraction and $\hat{n}$ is the refractive index.

Since the reflection coefficient is usually measured for unpolarized radiation, we have

$$\hat{R} = 0.5(|S_\||^2 + |S_\perp|^2). \qquad (1.9)$$

In addition, we also have the following formula for β:

$$\beta = \lambda \mu_t / 4\pi \qquad (1.10)$$

where μ_t is the linear absorption coefficient.

If we assume that the grazing angles are small, and substitute

$$\chi = \delta/\beta, \qquad x = \theta_1/\theta_c \qquad (1.11)$$

we obtain [16]

$$\hat{R}(\theta_1) = \frac{[\sqrt{2}x - (A + x^2 - 1)^{1/2}]^2 + A - (x^2 - 1)^{1/2}}{[\sqrt{2}x + (A + x^2 - 1)^{1/2}]^2 + A - (x^2 - 1)^{1/2}} \qquad (1.12)$$

where

$$A^2 = (x^2 - 1)^2 + (1/\chi)^2.$$

Equation (1.12) determines the family of parametric curves $\hat{R}(\theta_1)$ with $1/\chi$ as the parameter (figure 1.3) [14]. By constructing the plot of x as a function of $1/\chi$, and by choosing a suitable critical angle θ_c, we can uniquely determine x and χ and, consequently, the quantities $\hat{n}$ and β. Measurements of the optical constants, performed with the aid of (1.10)–(1.12), have produced very important experimental data that have been essential for the development of X-ray optical systems [16, 17]. The advent of multilayer X-ray polarizers [18] has meant that it is now possible to measure optical constants for different polarizations. The uncertainties in $\hat{n}$ and β determined in this way increase with increasing X-ray absorption because, for $\mu_t \geqslant 10\,\mu\mathrm{m}^{-1}$, the critical angle for total external reflection is subject to considerable uncertainty. Moreover, for $\lambda > 8\,\mathrm{nm}$, the angle of reflection exceeds $10°$ and the sine of an angle cannot be replaced by the angle itself (in radians), so that (1.12) ceases to be valid. The following more accurate formulae were therefore employed in [19] to calculate the optical constants:

$$\hat{R}_\perp = \frac{a^2 + b^2 - 2a\sin\theta + \sin^2\theta}{a^2 + b^2 + 2a\sin\theta + \sin^2\theta} \tag{1.13a}$$

$$\hat{R}_\parallel = \hat{R}_\perp \frac{a^2 + b^2 - 2a\cos\theta\cot\theta + \cos^2\theta\cot^2\theta}{a^2 + b^2 + 2a\cos\theta\cot\theta + \cos^2\theta\cot^2\theta} \tag{1.13b}$$

$$2a^2 = [(\hat{n}^2 - \beta^2 - \cos^2\theta)^2 + 4\hat{n}^2\beta^2]^2 + (\hat{n}^2 - \beta^2 - \cos^2\theta)$$

$$2b^2 = [(\hat{n}^2 - \beta^2 - \cos^2\theta)^2 + 4\hat{n}^2\beta^2]^2 - (\hat{n}^2 - \beta^2 - \cos^2\theta).$$

By measuring the reflection coefficients $\hat{R}_1(\theta_1, \delta, \beta)$ and $\hat{R}_2(\theta_2, \delta, \beta)$ at two angles of incidence, and by superimposing the corresponding $\delta\,(\beta)$ graphs, it is possible to determine δ and β from the points of intersection.

The graphical method of solving (1.13a) and (1.13b) provides a relatively accurate method of establishing the behaviour of the optical constants well away from absorption edges. However, the precision of the data obtained in this way deteriorates as we approach an absorption edge. The following procedure was used in [20, 21] to obtain quantitative data near absorption edges and also data on the anomalous dispersion of X-rays. The Fresnel formula for the complex reflection coefficient in the case of S-polarized radiation (electric field perpendicular to the plane of incidence) can be written in the form

$$S_\perp = \frac{a - \mathrm{i}b - \sin\theta}{a - \mathrm{i}b + \sin\theta} \tag{1.14}$$

where $a^2 - b^2 = \hat{n}^2 - \beta^2 - \cos\theta$, $ab = \hat{n}\beta$.

If we now write the complex reflection coefficient in the exponential form in terms of intensity reflection coefficient $\hat{R}$ and the phase shift ϕ of the wave

$$S = \sqrt{\hat{R}}\exp(\mathrm{i}\phi), \tag{1.15}$$

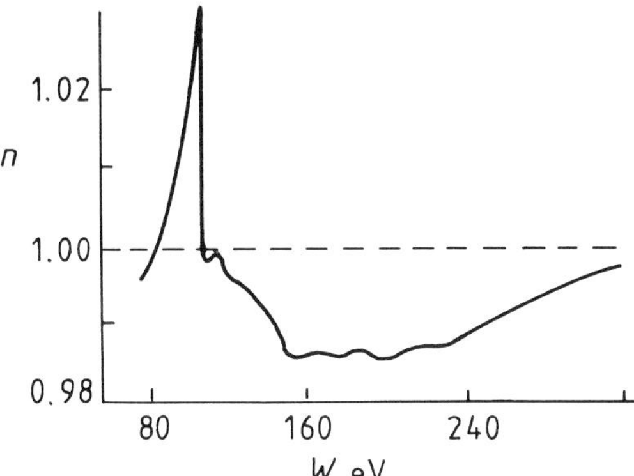

Figure 1.4 Real part of the refractive index of silicon as a function of photon energy [20].

we obtain the following expressions fo the coefficients a and b:

$$a = \frac{(1 - \hat{R})\sin\theta}{1 + \hat{R} - 2\sqrt{\hat{R}}\cos\theta}, \qquad b = \frac{-2\sqrt{\hat{R}}\sin\theta\sin\phi}{1 + \hat{R} - 2\sqrt{\hat{R}}\cos\phi}. \qquad (1.16)$$

The phase shift of the reflected wave can also be calculated from the reflection coefficient $\hat{R}$ as a function of the photon energy W_c, using the formula [22]

$$\phi(W_c) = \frac{1}{2\pi}\text{PV}\int_0^\infty \ln\left|\frac{W + W_c}{W - W_c}\right|\frac{\mathrm{d}[\ln\hat{R}(W)]}{\mathrm{d}W} \qquad (1.17)$$

where PV indicates that the principal value of the integral is to be taken. It is clear that a number of assumptions about the form of the function $\hat{R}(W)$ have to be introduced before the integral in (1.17) can be evaluated. The spectral function $\hat{R}(W)$ at given θ can be measured only within a limited energy range (W_c, W_2), so that the form of $\hat{R}(W)$ outside the range covered by the experimental values is of considerable importance for subsequent calculations. The expression

$$\hat{R}(W) = \hat{R}(W_2)(W_2/W)^m \qquad (1.18)$$

with $m = 2.5$ ($m = 4$ in the free electron model) was used in [20] for the purposes of extrapolation to shorter wavelegths ($W > 400\,\text{eV}$). A linear extrapolation was used at long wavelengths in the range to extrapolate the experimental curve to $\hat{R} = 1$ at $W = 0$ [20]. Figure 1.4 [20] shows the refractive index $\hat{n}$ as a function of the photon energy W for single-crystal silicon. The important feature of $\hat{n}(W)$ is the anomalous dispersion peak in the energy range 98–100.5 eV, in which δ changes rapidly and its magnitude reaches 1.003 at 100.5 eV. We shall see later that the presence of anomalous dispersion peaks can play a major part in the design of special optical elements in the case of 'white' synchrotron radiation.

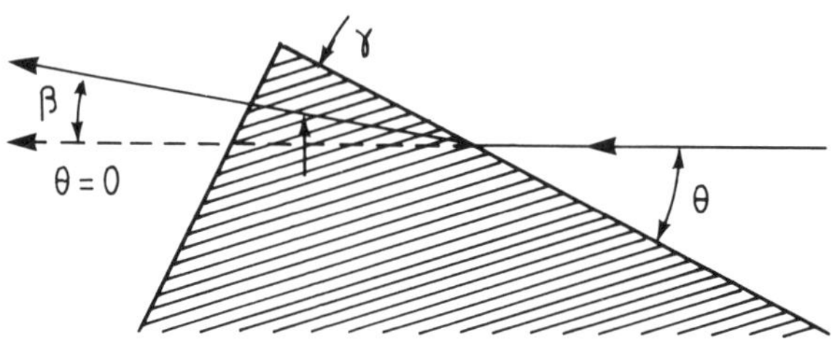

Figure 1.5   Prism method of measuring optical constants [26].

In addition to studies associated with measurements of the total reflection coefficient of crystal surfaces, there are also several publications dealing with applications of interferometric methods. The latter are very useful in studies of X-ray optical properties of materials with rough surfaces.

It is shown in [23] and [24] that measurement of the intensity spectrum for two or more transmission gratings can be used to obtain data on the optical properties of a given material. Under certain particular conditions, almost all materials from which gratings can be made are suitable for measurements of this kind. Moreover, most materials can be evaporated in vacuum on to a gold or platinum grating. The method does not depend on the incident-radiation intensity and can be used to monitor the process of deposition of different materials. For example, in the special case where a film is deposited on a grating with known optical constants, the unknown constants δ' and β' can be calculated from the expressions

$$
\cos 2\pi \left[\frac{t}{\lambda}\delta + \frac{t_1}{\lambda}\delta' \right] + \frac{Ip_1 - 1}{Ip_1 + 1} \cosh \left[2\pi \left(\frac{t}{\lambda}\beta + \frac{t_1}{\lambda}\beta' \right) \right] = 0
$$
$$
\cos 2\pi \left[\frac{t}{\lambda}\delta + \frac{t_2}{\lambda}\delta' \right] + \frac{Ip_2 - 1}{Ip_2 + 1} \cosh \left[2\pi \left(\frac{t}{\lambda}\beta + \frac{t_2}{\lambda}\beta' \right) \right] = 0
$$

(1.19)

where t is the thickness of the substrate grating, δ and β are its optical constants, t_1 and t_2 are the thicknesses of the deposited film at different times, δ' and β' are the constants of the material under investigation and $Ip_1 = [I_1^{(1)}/I_0^{(1)}](\pi^2/4)$ and $Ip_2 = [I_1^{(2)}/I_0^{(2)}](\pi^2/4)$ are the values obtained in the first and second measurements, respectively.

The greatest difficulties arise in the investigation of the X-ray properties of materials in the energy range near an absorption edge. Advances in absorption spectroscopy in the region of anomalous scattering have led to the creation of the method known as EXAFS (extended X-ray absorption fine structure), which is a spectroscopic technique for the precise determination of the properties of materials. Measurements of the real part of the atomic scattering factor in this

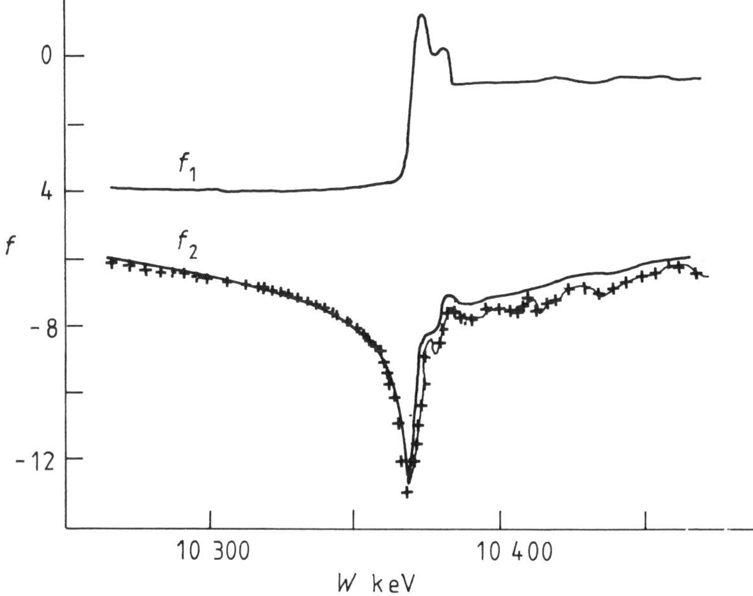

Figure 1.6 Measured atomic scattering factors f_1 and f_2 obtained with GaAs prism [26].

region are also very difficult. Accurate measurements of f_1 have been made by the interference method [25] which is suitable only for certain pure materials. The method described in [26], which relies on dispersion by a prism, has been tried on materials consisting of several components (elements).

Figure 1.5 illustrates the latter method [26]. The basic element is the prism made from the material under investigation. The incident beam enters the prism at an angle θ and is deflected by it as shown. It leaves in a near-normal direction, so that the overall deflection of the beam entering vacuum is small. The angle θ is only slightly greater than the critical angle θ_c for the material of the prism. Taking Snell's law in the form

$$\hat{n} \cos \gamma = \cos \theta \tag{1.20}$$

we find in the small-angle approximation that

$$\theta^2 - \gamma = 2\gamma = \theta_c^2.$$

If we know θ, we can determine δ by measuring the angle β.

This method was first used in 1924 and many of the results obtained in this way are summarized in [2]. The method was applied to the determination of the optical properties of gallium arsenide in [26]. Direct measurements of the deflection of a beam by a prism were used to find the refractive index and the absorption coefficient near the absorption edge of gallium (figure 1.6 [26]). Unfortunately, the prism method cannot be used at longer wavelengths because of much greater absorption.

A method employing multilayer interference coatings incorporating elements with different optical properties [27] has recently been added to existing interference methods for the determination of optical constants. Studies of the optical properties of materials in the X-ray range are continuing in many laboratories. The most complete data for a number of elements have been obtained on the DESY storage ring [28] under the direction of Professor Kunz. Data on the most widely used materials (boron and silicon) can also be found in [29, 30]. Organic films, widely used as substrates in filters and other X-ray optical elements, are studied to some extent in [31].

1.3 X-RAY OPTICAL ELEMENTS USING TOTAL EXTERNAL REFLECTION

The first X-ray optical elements exploiting the unique properties of materials in the nanometre wavelength range were introduced as far back as 1929 [32]. Total external reflection of X-rays incident on a polished surface at a small grazing angle is produced if the refractive index of the reflecting material is just less than unity. The principles underlying the fabrication and application of optical elements operating at small grazing angles have been described in a large number of publications, including the recent book by Michette [33].

We have already noted that total external reflection occurs at angles of incidence α close to $90°$ and is determined by the real part of of the complex refractive index $\hat{n}$ in accordance with the formula given by (1.7):

$$\theta_c = (2\delta)^{1/2}.$$

Table 1.2 lists the critical grazing angles for a number of elements at the indicated wavelengths [28].

It is clear from these data that the critical angles do not exceed a few degrees for wavelengths in the range 1–10 nm. The quality of the polished surface over a relatively large area is very important for such mirrors. The problem has already been discussed in early papers on X-ray optics. Another important effect is the astigmatism of spherical surfaces at small angles of reflection.

Ehrenberg [34] was the first to report the formation of images by plane and concave mirrors employing total external reflection, and demonstrated experimentally the effect of irregularities of the mirror surface on its focusing properties. Two-dimensional focused X-ray microimages were first demonstrated in 1948 by Kirkpatrick and Baez [35]. They removed astigmatism at grazing angles of incidence by employing two crossed cylindrical mirrors

Table 1.2

Element	λ, Å	θ_c°
C	44–280	2.6–23
	12–27	1.8–3.6
Mg	9.7–73	0.5–7.7
	302–1265	11.0–86
Al	206–990	7.3–86
Al$_2$O$_3$	12.4–90	1.8–12
Cu	2–41	2.6–8.9
Ag	7.7–15.5	2.6–4.1
	41.3–137.8	5.1–32
Au	4.1–15.5	2.6–4.4
	63.6–142.5	11–30

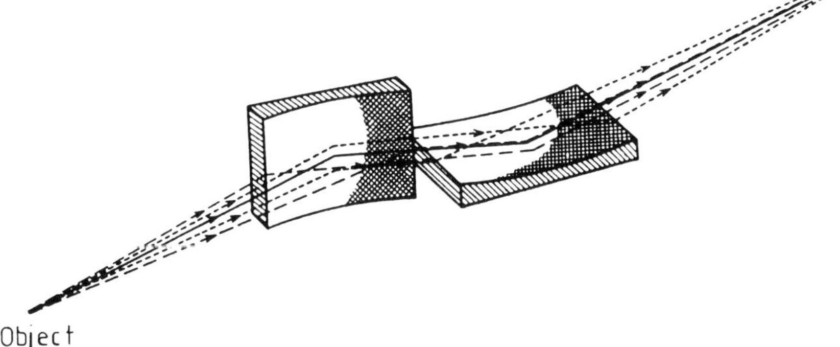

Figure 1.7 Principle of the Kirkpatrick–Baez microscope using total external reflection [35].

in their optical system, each of which was used to focus the X-rays in two orthogonal directions. The X-ray microscope of Kirkpatrick and Baez was a very successful instrument, widely used across the world [36]. Its optical system is illustrated in figure 1.7.

The resolution of the Kirkpatrick–Baez microscope used at present is close to the theoretical limit. The microscope with cylindrical mirrors can be optimized so that geometrical aberrations become comparable with diffraction effects [37].

McGee *et al* have shown that geometrical aberrations can be reduced by replacing cylindrical mirrors by surfaces described by a cubic equation [38]. The Kirkpatrick–Baez microscope is the main diagnostic tool in studies of plasmas produced in nuclear fusion systems [36, 39, 40].

The main disadvantage of the Kirkpatrick–Baez scheme is its small aperture angle. The axially symmetric arrangement of the Wolter microscope [41, 42] has a greater aperture. In the Wolter microscope of the first kind (figure 1.8), a hyperboloid and a paraboloid of revolution are combined so that they have a

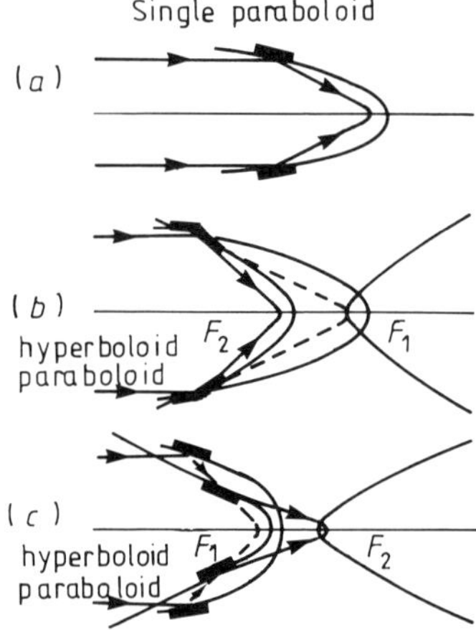

Figure 1.8 Wolter microscopes using total external reflection [41]: (a) scheme incorporating a paraboloid of revolution, (b) Wolter objective of the first kind, (c) Wolter objective of the second kind.

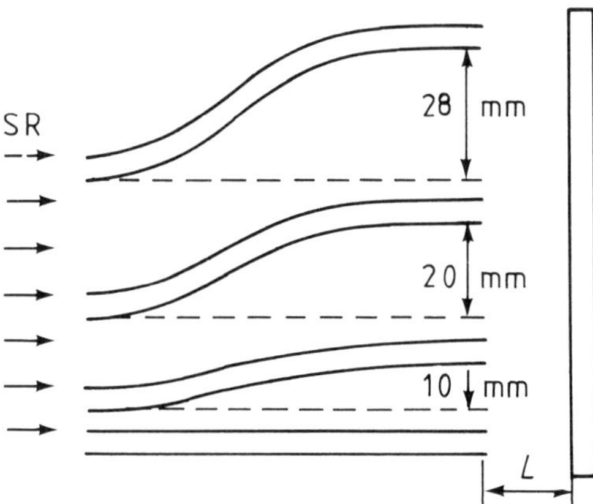

Figure 1.9 Experiment with a capillary waveguide [43]. The synchrotron radiation is incident from the left and the detector is at a distance L from the capillary.

common focus. In practice, the ellipsoid must have a minimum radius of 1–2 cm whereas the maximum radius can be a few metres or more. The precision with which the microscope elements must be fabricated is indicated by the following typical criteria: departure from calculated profile not more than 10 nm, surface irregularities not more than 1 nm, deviation from circular symmetry in a cross section not more than 0.5 μm. For a long time, the exceedingly stringent conditions imposed on the fabrication precision prevented the development of a

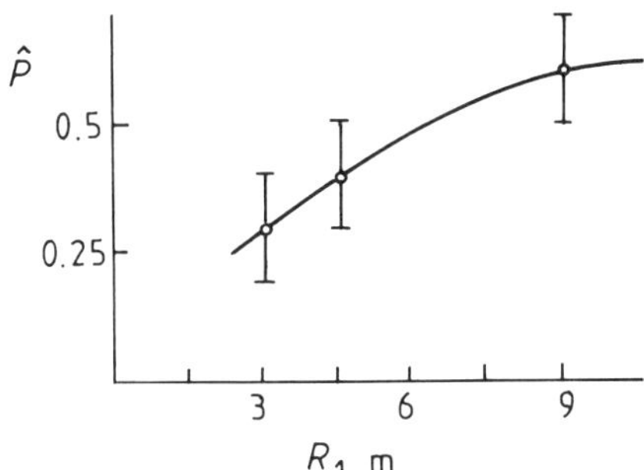

Figure 1.10   X-ray transmission coefficient of as a function of the radius of the capillary [43].

working prototype of this microscope. It was only a few years ago that a working Wolter microscope with a magnification of ×22, a field of view of 800 μm and a resolution of 4–5 μm was built at the Lawrence Livermore Laboratory [36]. The principle of reflection at grazing angles has recently found an original application in X-ray waveguides, i.e. X-ray optical fibres [43, 44].

Figure 1.9 illustrates an experiment on the transmission of X-rays in a capillary waveguide. The beam of synchrotron radiation is incident from the left on the capillary waveguide and is multiply reflected in the fibres with different radii of curvature. It then illuminates the detector on the right at a distance L from the waveguide. Figure 1.10 shows the measured transmission coefficient $\hat{P}$ as a function of the radius R of the capillary.

2

Planar diffraction optics

In this chapter, we shall describe the main properties of planar, i.e. two-dimensional, X-ray optical elements based on the principles of diffraction. We shall analyse the aberration properties of Fresnel zone plates and present the theory of operation. We shall compare different types of zone plate (amplitude-phase, holographic and kinoform) and will consider the formation of the focal spot for line zone plates. Finally, we shall present the results obtained in the course of fabrication and testing of phase zone plates based on silicon and designed for the wavelength range 0.7–2.0 nm.

We begin our survey of the principles underlying the operation of focusing diffraction optics with the arrangement shown in figure 2.1. This shows the interference pattern produced in the detecting medium when two spherical waves generated at points A_1 and A_2 are allowed to interfere. When the point A_1 is removed to infinity, the ellipsoids degenerate to paraboloids of revolution (figure 2.1(b)). Each curve in figure 2.1 represents a plane section, parallel to the optical axis, of a surface of constant phases that is an ellipsoid or paraboloid of revolution. All scattering points on a given ellipsoidal surface produce the same wave phase at A_2 if the source of the waves is at A_1. The phase difference between the ellipsoidal surfaces shown in figure 2.1 is chosen to be π (optical path difference $\Delta L = \lambda/2$).

We assume that a flat opaque screen K is placed between A_1 (source) and A_2 (image) and is perpendicular to the line $A_1 A_2$ (figure 2.2). A Fresnel zone plate is produced by cutting the screen so that only rings whose contributions to the image at A_2 have phases of the same sign are left open. The others remain opaque. The radius of the nth Fresnel zone (figure 2.2) is given by the expression (we neglect the second and higher powers of λ)

$$r_n = [nR_1 R_2 \lambda/(R_1 + R_2)]^{1/2} \tag{2.1}$$

where n is the order of the zone. The total aperture of the zone plate is $A = 2r_n$ where R_1 and R_2 are the corresponding distances from the screen K to the source and image, respectively. If we use the lens formula $1/R_1 + 1/R_2 = 1/F$, we can rewrite (2.1) in the form

$$r_n = (Fn\lambda)^{1/2}.$$

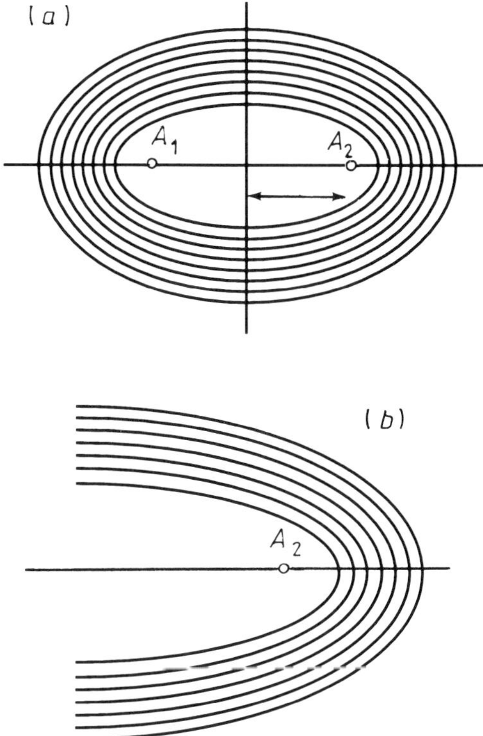

Figure 2.1   Interference pattern produced by two interfering spherical waves (a) and a plane and a spherical wave (b).

The intensity distribution in the focal plane is

$$I(r) = J_0^2(\pi A r)/4r^2 \tag{2.2}$$

where $J_0(x)$ is the zero-order Bessel function.

It will be useful to consider a few practical consequences of this formula, since they are used in calculations of the parameters of actual zone plates. The total aperture of a zone plate is given by

$$A = F\lambda/\delta r_n \tag{2.3}$$

where δr_n is the minimum width of the nth zone that can be fabricated by a given technology. It is important to recall that the minimum size of the focal spot of a zone plate is limited by a number of physical and technological factors. The lithographic process used in the fabrication of X-ray optical elements imposes the chief technological limitation. Physical limitations are imposed by the properties of materials such as their refractive indices and absorption coeficients.

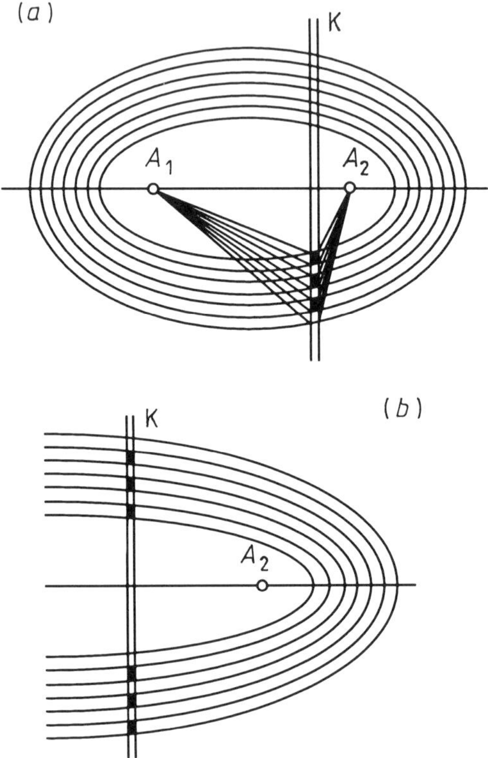

Figure 2.2   Arrangement used to produce elliptical (*a*) and parabolic (*b*) Fresnel zone plates.

2.1   PROPERTIES OF FRESNEL ZONE PLATES

The main properties of Fresnel zone plates are their resolving power, the relative aperture (relative amount of radiation collected by the plate in its focal spot) and the aberrations. Consider a section of a paraboloid of revolution, produced by cutting it with a flat screen at right angles to the optical axis (figure 2.2(*b*)). This case corresponds to the diffraction of a plane wave by a zone plate. The zone radii corresponding to an optical path difference of $n\lambda/2$ are given by

$$r_n = (n\lambda F + n^2\lambda^2/4)^{1/2}. \tag{2.4}$$

The optical properties of zone plates have been investigated by numerous workers [45–47]. For example, the intensity distribution in the diffraction pattern of a zone plate was investigated in [45, 46] where it was shown that, when the number of zones was large enough, the diffraction pattern on a plane perpendicular to the optical axis could be described by an Airy function. The intensity distribution along the optical axis has also been investigated. It was found that the focal-spot size along the axis decreased with increasing number of zones, and that there was a focal spot for each of the respective diffraction orders [48].

Next, we must introduce the concept of 'artificial zone plates'. These are fabricated by lithographic methods and usually have a rectangular line profile. 'Gabor zone plates' constitute another type and have a sinusoidal line profile. In the artificial plates, the covered zones consist of a perfectly opaque material and the fraction of energy collected in a diffraction order m is given by [48]

$$\frac{I_m}{I_p} = \frac{1}{[2(m-1)+1]\pi^2}.$$

(2.5)

It follows that, theoretically, the fraction of energy collected in first-order focal spot of an amplitude zone plate cannot exceed 10.1%. The focused wave combines in the focal plane with waves diffracted into other orders, and also those issuing from virtual foci, and this reduces the contrast of the focal spot.

The higher diffraction orders are absent in the case of the Gabor (sinusoidal) zone plate, but the fraction of radiation diffracted into the first order should not exceed 1/16, and the contrast problem remains [49]. The problem has been tackled [50, 51] by placing an opaque screen at the centre of the zone plate. This reduces the contribution of zero-order diffraction and substantially improves the contrast at the focal spot. The improvement in the signal-to-noise ratio is then described by the factor

$$\rho_n = 2n_c$$

(2.6)

where n_c is the number of zones covered by the screen. The optimum values of n_c lie in the range $10 < n_c < N/10$ where N is the total number of working zones.

The resolving power falls in the ratio $1 - n_c/N$ [50]. Zone plates with a central screen are known as 'apodized plates' and are widely used in X-ray microscopy. The contrast of an image produced by a zone plate is also determined by the ratio of the transmission coefficients of the light and dark zones.

The transmission function of a zone plate is described by

$$T(r) = \begin{cases} T_{\max} & \text{for even zones} \\ T_{\min} & \text{for odd zones} \end{cases}$$

(2.7a)

$$T(r) = T_0[1 + C_0 \sin(\pi r^2/\lambda F)]$$

(2.7b)

where

$$T_0 = T_{\max} + T_{\min}, \quad C_0 = (T_{\max} - T_{\min})/T_0.$$

The formula given by (2.7a) corresponds to the rectangular distribution of transmission, whereas (2.7b) is a continuous distribution (figure 2.3) [52]. The corresponding diffracted-wave amplitudes in the focal spot are given by [53]

$$E_F = NC_0T_0E_0\exp[\mathrm{i}2\pi F/\lambda]$$

(2.8a)

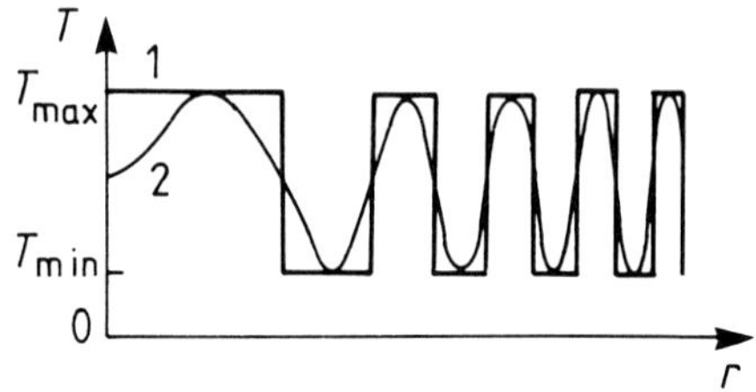

Figure 2.3 Transmission function of rectangular profile (1) and Gabor (2) zone plates [52].

and

$$E_F = (\pi/4)NC_0T_0E_0 \exp[i2\pi F/\lambda] \qquad (2.8b)$$

for rectangular and sinusoidal transmission distributions, respectively, where E_F is the wave amplitude in the focal spot and E_0 is the amplitude of the plane wave incident on the zone plate.

The radiation intensity at a focal spot is proportional to the difference of intensity transmission functions of light and dark zones:

$$|E_F|^2 \sim |T_{max} - T_{min}|^2.$$

For example, the focal-spot intensity can be reduced by 51% by depositing the zone plate on a Si_3N_4 membrane, 100 nm thick and having an intensity transmission factor of 0.8 (dark zones covered with gold film 0.14 μm thick with a transmission factor of 0.11).

The resolving power of the elements of the zone-plate optics is determined by diffraction and can be estimated from the Fresnel–Kirchhoff equations. The concept of the optical path length ΔL is useful in a qualitative solution of this problem. It is found that the field distribution in the focal plane can readily be obtained [53].

According to figure 2.4, the optical path difference is

$$\Delta L = AC - AC'.$$

Using the notation of figure 2.4, we have

$$\Delta L = r\beta_c \sin\theta - \frac{1}{2}F\beta_c^2$$

where we assume that $F \gg F\beta_c$ and that we can neglect the second term on the right-hand side of (2.4). According to Kirchhoff's diffraction theory, the total amplitude $E_{C'}$ at C' is equal to the sum of the contributions due to the individual points on the aperture of the optical element

$$E_{C'} = 2\int_0^{r_n}\int_{-\pi/2}^{\pi/2} E_0 \exp(-i2\pi \Delta L/\lambda)r\,dr\,d\theta. \qquad (2.9)$$

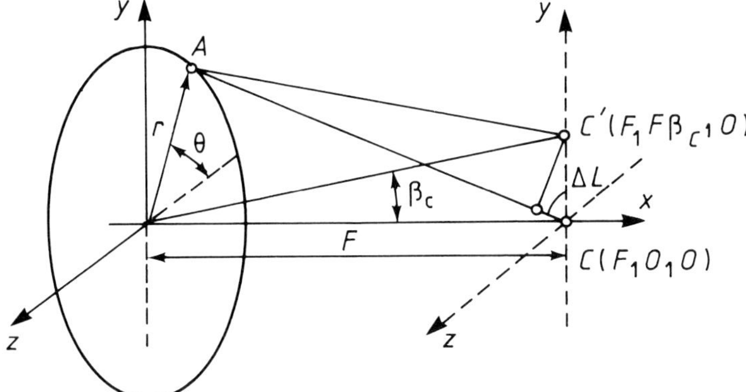

Figure 2.4 The diagram used to calculate the spatial resolution of a zone plate [53].

Substituting for ΔL in (2.9), and expressing the result in terms of zero-order Bessel functions, we obtain

$$E_{C'} = 2\pi E_0 \exp(i\pi F\beta_c^2/\lambda) \int_0^{r_n} J_0(2\pi\beta_c r/\lambda)r\,dr. \qquad (2.10)$$

In practice, the integral in (2.10) is evaluated over the unobstructed zones with radii given by (2.4). It is shown in [53] that the amplitude zone plate with only the negative zones obstructed (from r_{2n+1} to $r_{2(n+2)}$) has better resolution than the plate with the same parameters but with unobstructed positive zones (from r_{2n} to r_{2n+1}). Actually, according to (2.10), the radiation intensities $I_{C'}^+$ and $I_{C'}^-$ at C' due to positive and negative zones are given by the following expressions (after evaluating (2.10)):

$$I_{C'}^+ = \frac{16}{(N+1)^2}\left\{ \sum_{n=0}^{(N-1)/2} \frac{1}{\rho_a}\left[(2n+1)^{1/2}J_1\left[\rho_a(2n+1)^{1/2}\right] \right.\right.$$
$$\left.\left. - (2n)^{1/2}J_1\left[\rho_a(2n)^{1/2}\right] \right]\right\}^2 \qquad (2.11a)$$

$$I_{C'}^- = \frac{16}{N^2}\left\{ \sum_{n=0}^{(N-1)/2} \frac{1}{\rho_a}\left[[2(n+1)]^{1/2}J_1\left[\rho_a[2(n+1)]^{1/2}\right] \right.\right.$$
$$\left.\left. - (2n+1)^{1/2}J_1\left[\rho_a(2n+1)^{1/2}\right] \right]\right\}^2 \qquad (2.11b)$$

where N is the total number of zones and $\rho_a = 2(\pi\beta_c/\lambda)(\lambda F)^{1/2}$.

The intensity distribution in the focal plane is very similar to that for a lens, i.e. $J_1(r)/r$, but the positions of the first zero and the first oscillation (second order) are significantly different. If we take P_0 to be the value of the argument at zero intensity (first minimum of the distribution), we find that

$$\beta_c A/\lambda = P_0 N^{1/2}/\pi. \qquad (2.12)$$

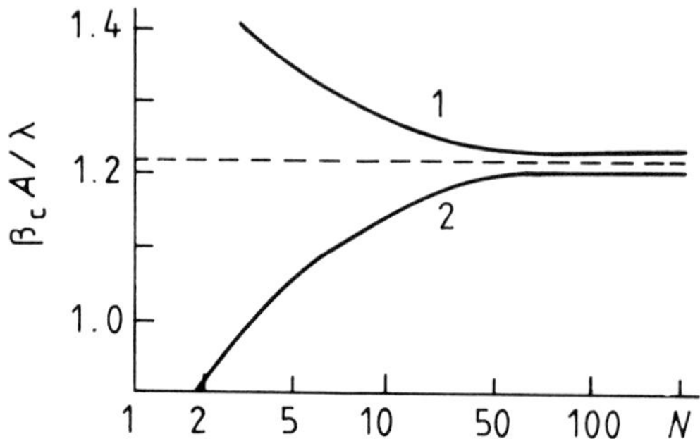

Figure 2.5 Resolution of positive (1) and negative (2) zone plates as a function of the number of zones.

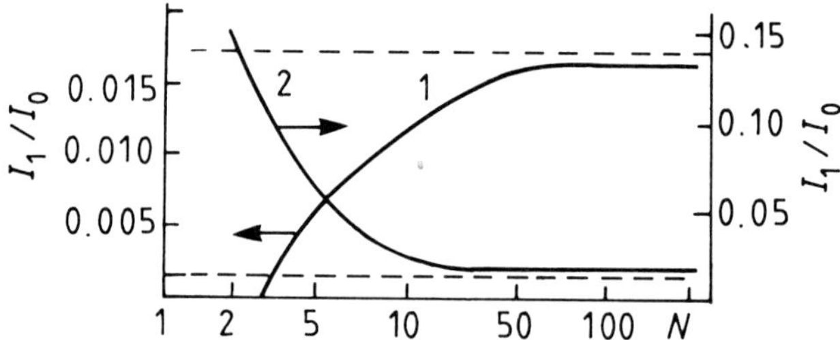

Figure 2.6 Relative intensity as a function of the number of zones for positive (1) and negative (2) plates, compared with ordinary lenses of the same aperture.

Figure 2.5 shows the quantity $\beta_c A/\lambda$ as a function of the number of zones in the case of positive and negative zone plates. The dashed line shows the spot size due to a lens of diameter A.

The resolution of a negative zone plate is therefore worse than that of an ordinary lens of the corresponding diameter. It is only when the zone number becomes 200 or more that the resolution of the simple lens is approached asymptotically by both positive and negative zone plates. However, although negative zone plates have better resolution than simple lenses, they produce lower contrast than both simple lenses and positive zone plates. Figure 2.6 shows the intensities of the first diffraction orders calculated from (2.11a) and (2.11b), and also those due to a simple lens (dashed lines). The high intensity obtained for the negative plate at low N produces a sharp reduction in the signal-to-noise ratio. It is often better to use zone plates with a relatively large number of zones ($N = 20$–100).

2.1.1 Aberrations of zone plates

The aberrations of zone-plate elements are conveniently analysed in terms of the optical path difference ΔL. Consider the classical parabolic zone plate described by (2.4). The plate is aberration free when it is illuminated by a monochromatic plane wave propagating along the optical axis. The optical path difference is $\lambda/2$ and the focal-spot size is limited by the aperture of the zone plate alone. Aberrations arise when the zone plate is used to image an object placed at a finite distance from it, or when the object is in an off-axis position, or when non-monochromatic radiation is employed. For a point object lying on the optical axis at a finite distance R_1 from the zone plate, the optical path difference is given by

$$\Delta L = (R_1^2 + r_n^2)^{1/2} + (R_2^2 + r_n^2)^{1/2} - (R_1 + R_2) \tag{2.13}$$

where R_2 is the distance from the lens to the image and r_n is the radius of the nth zone.

(a) *Spherical aberration*
Combining the equation for ΔL with (2.4), we find that, to within second-order terms in λ,

$$\Delta L = \frac{1}{2}n\lambda + \frac{1}{8}n^2\lambda^2\left[1 - \frac{R_1^3 + R_2^3}{(R_1 + R_2)^3}\right] \tag{2.14}$$

where we have used the approximate expansion

$$(1 + x^2)^{1/2} = 1 + x^2/2 - x^4/8.$$

The second term in (2.14) represents to the spherical aberration of the zone plate. Using the lens formula, we can find the maximum of the aberration term by equating to zero the derivative of (2.14) with respect to R_1. When this is done, we find that the spherical aberration is a maximum for $R_1 = 2F$, i.e. when the image is transmitted with unit magnification. Since aberrations become appreciable when the aberration term becomes comparable with $\lambda/4$, we then find that

$$N^2 < 8F/3\lambda. \tag{2.15}$$

When the wavelength is $\lambda = 4.5\,\text{nm}$ and the focal length is $10\,\text{mm}$, the maximum number of zones producing an aberration-free image is found to be 2434. The spherical aberration of zone plates is relatively insignificant at X-ray wavelengths, but it increases as longer wavelengths are approached. For vacuum ultraviolet radiation generated by an excimer laser ($\lambda = 193\,\text{nm}$), the number of zones of an aberration-free zone plate with the same focal length is 371. Aberrations can be avoided altogether by using an ellipsoidal zone plate, specially designed to image objects located at finite distances from the plate. The zone-plate configuration is then obtained by cutting the surfaces of equal phase (ellipsoids of revolution) with a plane perpendicular to the optical axis (figure 2.2(a)). This type of zone configuration is best described in terms of ellipsoidal coordinates [54]. The coordinate surfaces in this system are confocal ellipsoids and paraboloids of revolution. If we confine our attention to the two-dimensional case, we find that Cartesian coordinates transform into the ellipsoidal form in accordance with the expressions

$$x = a\xi\eta \qquad y^2 = a^2(\xi^2 - 1)(1 - \eta^2) \tag{2.16}$$

where $\xi \geqslant 1$, $-1 < \eta < 1$ and a is the focal parameter (figure 2.1(a)). Ellipsoidal coordinates and their properties will be discussed in detail in chapter 4. Since

$$R_1 + R_2 = 2a + \frac{1}{2}n\lambda \tag{2.17}$$

the equation of the plane perpendicular to the optical axis is

$$a\xi\eta = a - R_2 \tag{2.18}$$

and the expression for the parameter of confocal ellipsoids of revolution corresponding to a phase diference of $\lambda/2$ is

$$\xi_n = 1 + n\lambda/4a \tag{2.19}$$

the radius of the nth Fresnel zone is given by

$$r_n^2 = \left[an\lambda/2 + n^2\lambda^2/16\right]\left[1 - (a - R_2)^2/(a + n\lambda/4)^2\right]. \tag{2.20}$$

The denominator terms in the second factor become significant only when focal-spot size is comparable with the wavelength λ. It is important to note, however, that (2.20) is the exact expression for the zone radii, which ensures the absence of aberrations. This expression becomes identical with (2.4) as $a \to \infty$.

(b) *Chromatic aberrations*
The focal length of a given zone plate is a function of wavelength, so that chromatic aberrations become very considerable when the plate is illuminated with non-monochromatic radiation. Actually, if the optical path difference is exactly $n\lambda/2$, chromatic aberrations are significant for

$$\Delta L = n\lambda/2 \pm \lambda/4 \tag{2.21}$$

so that the maximum number of zones for which chromatic aberration can still be neglected is given by

$$N \approx \lambda/\Delta\lambda \tag{2.22}$$

where $\Delta\lambda$ is the full width at half height of the incident spectrum.

The expression given by (2.22) is used in estimates of the resolving power of zone plates. For example, for $N = 100$ and $\lambda = 4.5\,\text{nm}$, we find that $\delta\lambda$ should not exceed $0.045\,\text{nm}$. The chromatic properties of zone plates have often been used to select the spectrum of transmitted radiation [50, 55, 56]. Condenser zone plates made by the holographic method are particularly suitable for such

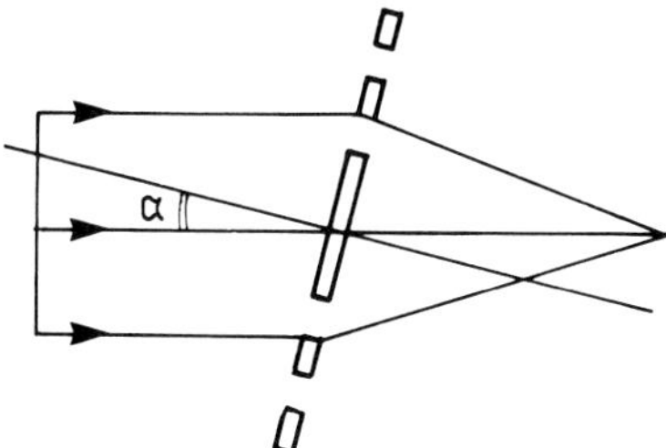

Figure 2.7   Inclined incidence of radiation on a zone plate [59].

purposes. A specially constructed optical system was used in [57] to produce zone plates with zone number up to 1196 and last zone width of 0.458 μm. At $\lambda = 4.5$ nm, the spectral resolving power of this plate is $\lambda/\Delta\lambda \approx 250$. An image resolution of 60 nm was achieved in [58] by using a combination of a condenser zone plate, a monochromator, a microzone plate and an objective.

(c) Off-axis aberrations

These aberrations arise when a zone plate is illuminated by a plane or spherical wave from an off-axis source. The case of an inclined objective wave is illustrated in figure 2.7. The expression for the optical path difference in the case of a plane wave then takes the form [59]

$$\Delta L = r_n^3\alpha/2F^2 - 3r_n^2\alpha^2/4F + n\lambda/2. \tag{2.23}$$

It is important to note that (2.23) contains only terms corresponding to the Seidel aberrations, but there are no terms representing field distortion—the principal aberration of optical lens systems.

The Fresnel zone plate produces an image that is totally free from distortion in the focal plane. The first term in (2.23) describes coma and the second represents astigmatism and field curvature. Aberrations become significant when the sum of the terms becomes equal to $\lambda/4$. For a large number of zones, the aberration-free field is quite small. The situation is then dominated by coma, and the angle α of an aberration-free field of view is

$$\alpha = (N\lambda/F)^{-1/2}(2N)^{-1}. \tag{2.4}$$

For example, when $\lambda = 4.5$ nm, $N = 1000$ and $F = 10$ mm, the angular field of view (the half-angle) is 1.35°. For a smaller number of zones, e.g. for the microzone plates of X-ray microscopes, in which the number of zones does not exceed 100, the principal aberrations are astigmatism and field curvature, and we have

$$\alpha \approx 1/(3N)^{1/2}. \tag{2.25}$$

For $\lambda = 4.5\,\text{nm}$, $N = 100$ and $F = 10\,\text{mm}$, the aberration-free image is observed for $\alpha = 3.3°$, i.e. for a field of $500\,\mu\text{m}$, which is much larger than the aperture of the zone plate itself.

Finally, consider the case where the object is at a finite distance from the zone plate. If we expand the expression for the optical path difference into a power series, we again find that there is no field distortion, and the remaining aberrations are given by the following expressions:

$$\text{coma:} \qquad \cos\theta\,(r_2^3\alpha/2)(1/R_2^1 - 1/R_2^2) \qquad (2.26a)$$

$$\text{astigmatism:} \qquad -[(r_n^2\alpha^2/(2F)]\cos^2\phi \qquad (2.26b)$$

$$\text{field curvature:} \qquad -[r_n^2\alpha^2/(2F)]. \qquad (2.26c)$$

The factor $\cos\theta$ is introduced to represent the two-dimensional distribution of points in the plane of the object. The image depth of field due to a zone plate is given by [60]

$$\Delta F = \lambda F^2/(4r_n^2). \qquad (2.27)$$

For example, a plate with 250 zones and a focal length of $1.3\,\text{mm}$ has a depth of field $\Delta F = 1.3\,\mu\text{m}$ [60]. The spatial resolution of this zone plate is $50\,\text{nm}$.

These formulae describe the aberration properties of the zone plate as an image-forming device. They can be used to estimate image quality in particular cases. The absence of field distortion is an important feature. Together with the high spatial resolution, this has led to the utilization of zone plates in X-ray microscopy [58]. However, at present, Fresnel zone-plate micro-objectives fall far short of the ideal in two important respects, namely, relative aperture and efficiency.

2.1.2   The relative aperture and efficiency of zone optics

The relative aperture of a zone plate is determined by the maximum resolution of the fabrication process. For a microzone plate with a resolution of $70\,\text{nm}$, the ratio of the area of the entrance stop to the square of the focal length does not exceed 1.5×10^{-3}. This figure is difficult to improve upon because it depends on the width of the outermost zone. As far as the efficiency of zone plates is concerned, we have already noted that the efficiency of traditional amplitude zone plates does not exceed 10%. The efficiency problem can be partially solved for the amplitude zone plate by using a composite plate [61] in which a central zone plate of first-order focal length F is surounded by additional zones with an equal third-order focal length. The zone radii of this type of plate are given by [61]

$$r_n^2 = n\lambda F \qquad n = 1, 2, \ldots, N_1 \qquad (2.28a)$$

$$r_n^2 = n\lambda F(3 - 2N_1/M) \qquad n = N_1 + 1, N_1 + 2, \ldots, M. \qquad (2.28b)$$

Figure 2.8 shows the general appearance of the composite zone plate together with radial intensity distribution in its focus for a plate in which the number of

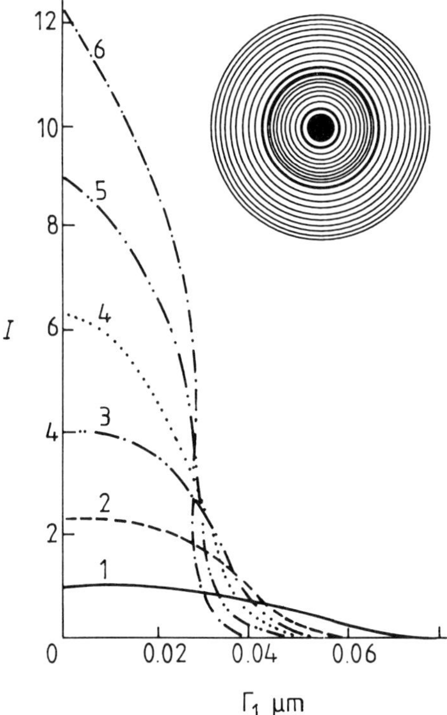

Figure 2.8 General appearance of a composite Fresnel zone plate and the radial intensity distribution in its focal plane for $N_1 = 100$ for number of outerzones M equal to 100 (1), 150 (2), 200 (3), 250 (4), 300 (5) and 350 (6) [61].

zones is $N_1 = 100$. Composite zone plates are also found to have improved resolution because the overall aperture of the optical element is increased by a maximum factor of two. The main advantage of the composite plate is the higher intensity it produces in the focal spot. The focal-spot intensity can in fact be increased by more than an order of magnitude in comparison with an ordinary plate of comparable resolution by increasing the number of zones to 350. The best parameters of a composite zone plate are achieved for $M = 11N_1/3$. When this condition is met, the potentialities of the fabrication technology are exploited to the full if the outermost zones of the inner and outer structures have the same width. In the optimum case, the focal-spot intensity can thus be increased by a factor of 13.4. However, the integral efficiency of the composite plate (the amount of energy diffracted into the focal spot) remains small. According to (2.28b), the total aperture of the optimum composite zone plate is $9N_1$. This means that, according to (2.28a), the focal-spot intensity is higher by a factor of 81, whereas according to [61] the intensity rises by a factor of only 13.4. Numerical calculations show that the total efficiency amounts to only 1.5%.

(a) *Phase zone plates*

The above properties of zone plates made of absorbing materials suffer from an important defect: the efficiency of the collection of radiation into the focal spot is relatively low (theoretically, not more than $1/\pi^2$); moreover, a quarter of the incident radiation is not diffracted and contributes to the strong background in the image plane. A way out of this difficulty was proposed 1888 by Rayleigh who put forward the idea of the phase zone plate [62] and Wood then demonstrated the advantages of these plates as compared with amplitude plates [63]. The Rayleigh–Wood devices have been very useful in X-ray and vacuum ultraviolet optics in which there are no analogues of ordinary optical elements. In the Rayleigh phase-contrast plate, all the Fresnel zones are clear and transparent to radiation. Focusing is achieved by ensuring that alternate zones introduce a phase shift such that the interfering waves in the focal spot all have phases of the same sign.

Consider a plane wave incident on a zone plate (figure 2.2(b)). Amplitudes from different zones are combined in the first-order focus with different phases. The optical path length to the focus increases with increasing radial distance on the zone plate. The phase increases from 0 to π within the first zone (r_0-r_1), and the phase of the average amplitude is $\pi/2$. Within the second zone (r_1-r_2), the phase rises from π to 2π, and the average phase is $3\pi/2$. The resultant wave amplitude due to the first and second zones is therefore zero, since the interfering waves are in antiphase. However, if the first zone introduces an extra phase of π, the waves interfere constructively in the focal spot, and similarly for all the other zones for which the phase shift is greater than 2π.

The resolution and aberrations of the Rayleigh zone plate are entirely analogous to those of the amplitude plates [64, 65]. Table 2.1 compares amplitude and phase plates in terms of the distribution of diffracted intensity (I_1 is the first-order intensity and I_2, I_3, ... are the intensities in the second, third, *etc.* diffraction orders) [64].

We have reproduced the diffracted intensities in the ideal case of completely transparent and completely opaque materials. In practice, the zone plate material introduces a phase shift and also absorbs. Table 2.1 also lists the results for the kinoform zone plate that has a special surface pofile. The properties and structure of kinoform plates will be discussed later.

It is clear from our review of the optical properties of materials in the soft X-ray region, presented in chapter 1, that all materials exhibit considerable absorption in this range. It is therefore impossible to fabricate the ideal Rayleigh–Wood phase-contrast plate. A practical structure constitutes a more or less satisfactory approximation to the ideal structure. Its diffracting properties are determined by the complex refractive index $\hat{n}$. A practical phase zone plate, with properties that are much better than those of the amplitude plate, can be produced by carefully selecting the plate material. The basic quantity that affects the phase modulation of the plate structure is the ratio

Table 2.1

Type of zone plate	I_1/I_0, %	I_2/I_0, I_3/I_0, ..., %	Undiffracted portion, %	Absorption, %
Fresnel (amplitude)	10.1	0, 1.1, 0, 0.4, ...	25	50
Rayleigh–Wood (phase)	40.4	0, 4.5, 0, 1.6, ...	0	0
Gabor (amplitude)	6.25	0	25	68.75
Gabor (phase)	34	10, 1, 0.1, 0.06, ...	0	0
Kinoform (phase)	100	0	0	0

$$\chi = \beta/\delta. \tag{2.29}$$

For materials consisting of only one element (from the Periodic Table), this ratio is

$$\chi = f_2/f_1. \tag{2.30}$$

If $\beta = 0$ and, consequently, $\chi = 0$, the material is perfectly transparent. The thickness of the phase-shifting layer that will ensure maxium efficiency of the zone plate is obviously given by

$$t_\pi = \lambda/2\delta. \tag{2.31}$$

This thickness difference produces a phase shift of π radians. In practice, the maximum focal-spot intensity is achieved for somewhat different layer thicknesses [64].

Consider a zone plate in which odd zones are unobstructed and even zones are covered by a phase-shifting material (positive zone plate). The attenuation of amplitude and the phase shift produced by the even zones are then given by

$$\Delta\varphi = 2\pi\delta t/\lambda \tag{2.32}$$

$$\Delta\beta = -2\pi\beta t/\lambda - \chi\Delta\varphi. \tag{2.33}$$

The amplitude of the wave transmitted by an unobtructed zone is given by

$$E_1 = \frac{E_0}{2\pi} \int_0^\pi \exp(i\varphi)d\varphi = \frac{iE_0}{\pi} \tag{2.34}$$

where E_0 is the amplitude of the incident wave. The amplitude of the wave transmitted by a covered zone is

$$E_2 = \frac{iE_0}{\pi} \exp(-\chi\Delta\varphi - i\Delta\varphi). \tag{2.35}$$

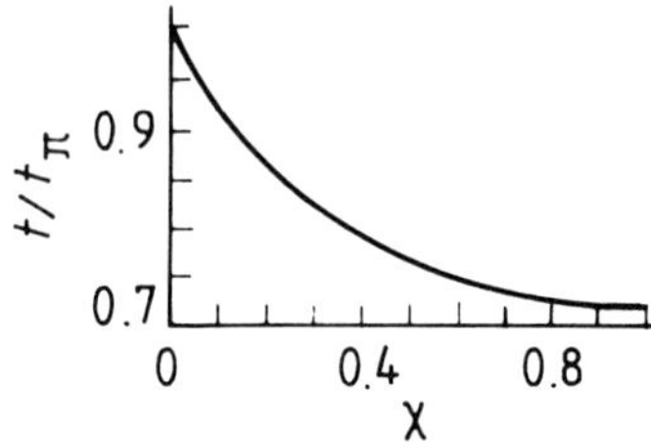

Figure 2.9   Relative thickness of a zone plate.

Consequently, the total first-order intensity transmitted by the first and second zones is given by

$$I_1^2 = \frac{E_0^2}{\pi^2}[1 + \exp(-2\chi\,\Delta\varphi) - 2\cos\Delta\varphi\,\exp(-\chi\,\Delta\varphi)]. \tag{2.36}$$

The contribution of higher orders is readily taken into account. It can be calculated from the following formulae:

$$I_1 = \begin{cases} E_0^2[1 + \exp(-2\chi\,\Delta\varphi) - 2\cos\Delta\varphi\,\exp(-\chi\,\Delta\varphi]/(m^2\pi^2)] & m\text{ odd} \\ 0 & m\text{ even.} \end{cases} \tag{2.37}$$

Table 2.1 lists the numerical results obtained by using (2.37) with $\chi = 0$ and $\Delta\varphi = \pi$. If there is appreciable absorption by the material, the optimum phase shift and, consequently, the optimum layer thickness, can be found from the expression

$$2\sin\Delta\varphi_{\mathrm{opt}} + 2\chi\cos\Delta\varphi_{\mathrm{opt}} = 2\chi\,\exp(-\chi\,\Delta\varphi_{\mathrm{opt}}). \tag{2.38}$$

The optimum phase shift $\Delta\varphi_{\mathrm{opt}}$ can be used to calculate the optimum thickness of the amplitude-phase plate that will ensure the maximum efficiency in the focal plane [64]:

$$t_{\mathrm{opt}} = \lambda\,\Delta\varphi_{\mathrm{opt}}/2\pi\delta. \tag{2.39}$$

Figure 2.9 shows the optimum thickness of a zone plate as a function of the parameter χ in units of t_π. Once the corresponding χ and $\Delta\varphi_{\mathrm{opt}}$ have been chosen, we can readily estimate the fraction of energy absorbed by the zone plate or diffracted into the zero order. The absorbed intensity is

$$I_{\mathrm{abs}} = \frac{E_0^2}{2}[1 - \cos(-2\chi\,\Delta\varphi)] \tag{2.40}$$

and the intensity of undiffracted radiation is

$$I_0 = \frac{E_0^2}{4}[1 + \exp(-2\chi\,\Delta\varphi) + 2\cos\varphi\,\exp(-\chi\,\Delta\varphi)]. \tag{2.41}$$

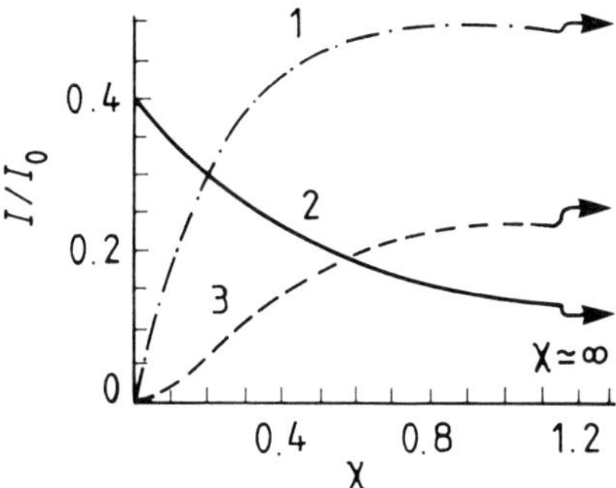

Figure 2.10 Relative absorbed (1), focused (2) and undiffracted (3) radiation as functions of χ.

Figure 2.10 [64] shows the relative absorbed energy and the energy of focused and undiffracted radiation for the optimum zone-plate thickness and different values of χ. Estimates based on (2.36) show that the phase properties of zone plates are well defined for $\chi < 0.175$. For such values of χ, the expression given by (2.36) can be simplified by taking $t_{\mathrm{opt}} \sim t_\pi$ so that [66]

$$E_1^2 = E_0^2[1 + \exp(-\chi\pi)]^2/\pi^2.$$

Knowledge of χ alone is not sufficient for the design of a zone plate. The numerical value of t_{opt} is also important from a technological point of view. When $t_{\mathrm{opt}} \geqslant 1\,\mu$m, zone plates are rather difficult to fabricate by the methods of microelectronic technology [66]. There are some materials, however, for example, silicon, where technology is available for producing structures with a very large height to width ratio (10 to 15). Figure 2.11 [67] show photographs of zone plates made from single-crystal silicon by electron-beam lithography and ion-chemical etching. The parameters of these plates were as follows:

Total number of Fresnel zones	56
External diameter, μm	114
Width of outermost zone, nm	300
Thickness of phase-shifting layer, μm	2.4
Plate material	Silicon
Focal length ($\lambda = 0.834\,$nm), cm	5

For silicon and $\lambda = 0.834\,$nm, we have $\chi = 0.045$ and $t_{\mathrm{opt}} = 2.3\,\mu$m. The characteristics of the zone plate of figure 2.11(b) are close to those of the ideal characteristic. The calculated efficiency is 35% [67–69].

Figure 2.12 shows the calculated efficiency of a silicon zone plate and the thickness of the phase-shifting layer as functions of wavelength in the range 0.6–4 nm. These graphs can be used to choose the range in which silicon is a

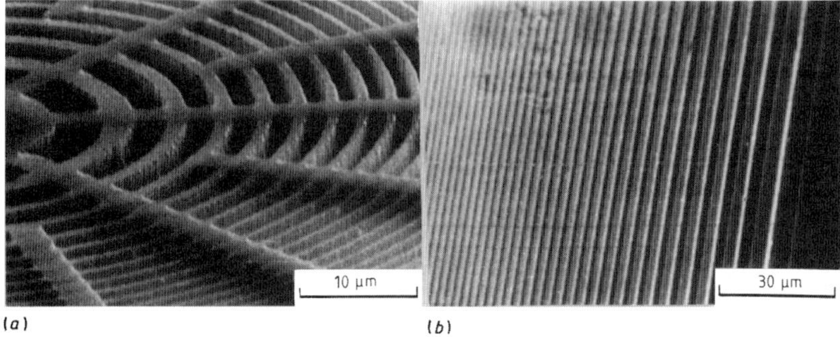

Figure 2.11   Concentric (*a*) and linear (*b*) zone plates made from silicon [67].

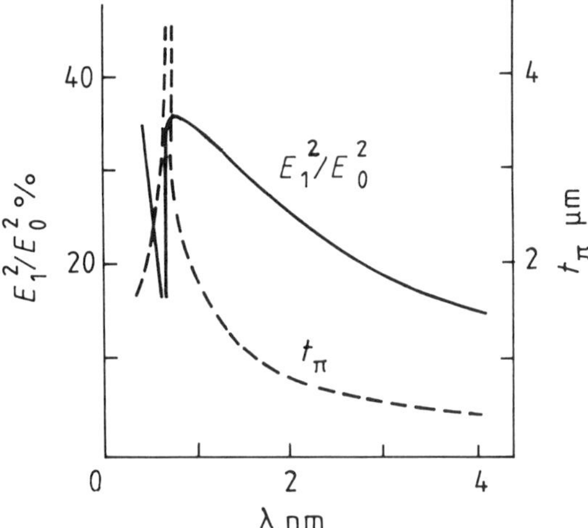

Figure 2.12   Eficiency and thickness t_π as functions of wavelength for a silicon zone plate [67].

satisfactory material for X-ray optics. Naturally, any departure of the thickness of the phase-shifting layer from the optimum value results in lower diffraction efficiency of the zone plate. However, such variations in thickness can be allowed to lie between relatively wide limits. For example, a 20% change in the thickness of this layer produces only a 5% reduction in the intensity in first-order focus. In the case of silicon, noise with an amplitude amounting to 1% of the total intensity in the focal spot is produced when the layer thickness is allowed to vary by ±10%, i.e. ±0.23 μm. This degree of surface roughness does not impose particularly stringent conditions on surface polishing.

Silicon zone plates have been tested with characteristic radiation from an X-ray tube incorporating an aluminium anode (λ = 0.834 nm). The measurements were made in a vacuum because radiation of this wavelength is readily absorbed

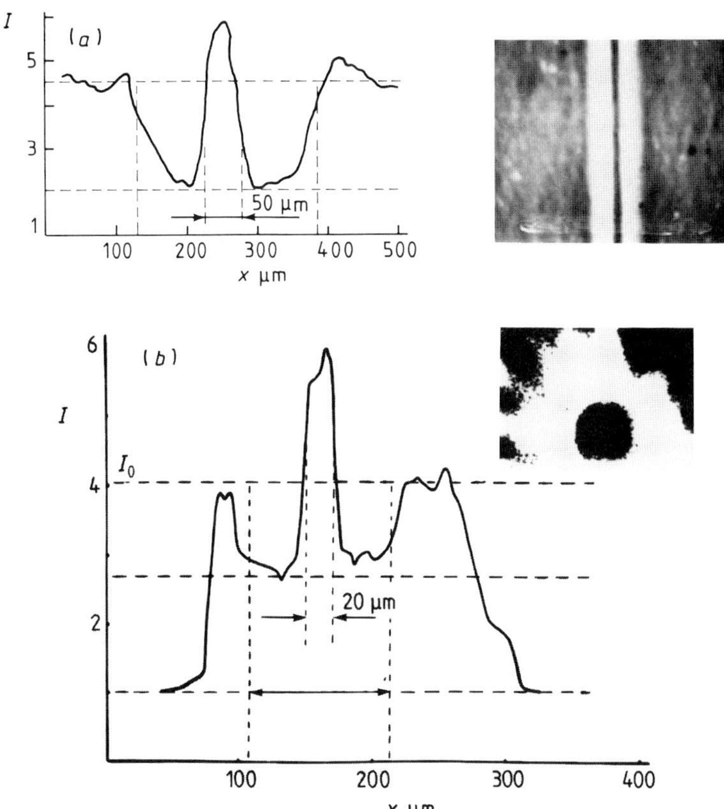

Figure 2.13 Energy distribution in the image plane for linear (*a*) and concentric (*b*) zone plates.

by air. An aluminium filter placed between the source and the object slit was used to isolate the characteristic X-ray line. The radiation was recorded on plates with an intrinsic resolution of 1000 lines mm^{-1}. Figure 2.13 shows images recorded in the focal plane of zone plates, together with the corresponding microdensitometer scans. The efficiency of the concentric zone plate was estimated from these data as 18%. The focal-spot size was determined by the optical system used in this experiment.

Quantitative measurements of the efficiency of these zone plates were made by calibrating the photographic plates on a densitometer (at $\lambda = 0.834\,\text{nm}$). The measured contrast factor was $\Gamma \approx 1.12$.

Apart from silicon, there is a whole series of other materials that are suitable for phase zone plates [64, 66]. Table 2.2 lists some of these materials, selected

Table 2.2

Material	λ nm	Material	λ nm
C	5.1–8.6	Au	2.0–2.34
Al	1.45–2.25	Nb	0.7–0.8
Si	0.7–2.05	Mo	0.7–0.8
Cu	0.44–0.54		6.4–40
Ag	0.46–0.7	PMMA	5.0–8.0
	7.2–11.0	Polyimide	5.9–8.0

Table 2.3

Material	Absorption edge	λ nm	$-\chi$	t_{opt} nm	I_1/I_0 %
Be	K	11.27	0.05	0.71	34.9
Si	L	12.4	0.048	0.2	35
Te	M_5	21.7	0.1	0.2	30

according to the following principles. The calculated efficiency of phase zone plates fabricated from these materials should be in excess of 25% and the optimum thickness of the phase-shifting layer should be less than or of the order of 1 μm, which is at the limits of current technology. In addition to silicon plates [67], the fabrication of germanium [70], silver [66] and carbon [71] plates has been reported.

Tests on the condenser zone plate made from germanium have demonstrated very good agreement with calculations based on (2.36) [70]. The first-order diffraction efficiency measured in [70] was 15% (taking into account the absorption of radiation in the supporting membrane).

Interesting possibilities emerge when single-element media are used at wavelengths corresponding to anomalous absorption and refraction. In the previous chapter we gave some examples of the change in the optical properties of materials in the region of anomalous scattering. Unfortunately, there is a lack of data on the optical constants of many materials in these ranges. However, such data do exist for a few materials, e.g. silicon [20, 21].

A zone plate operating in the anomalous-scattering region must have good wavelegth selective properties and should be capable of producing an image in 'white' radiation.

Table 2.3 lists some of the constants of materials that are good candidates for this kind of 'polychromatic' optics that exploits the change in the sign of the real part of the optical constant (according to the data given in [5]).

Figure 1.4 shows the calculated refractive index of silicon in the region of the L-absorption edge. It has well defined anomalous properties. The real part of the refractive index become positive at about 100 eV. An 'anomalous lens' (after Yu A Basov) can also be used as a natural monochromator. The refractive index undergoes an anomalous change within a wavelength interval of width $\Delta\lambda/\lambda = 0.01$ (figure 1.4).

2.2 HOLOGRAPHIC ZONE PLATES

The holographic method is widely used to produce the so-called Gabor or holographic zone plates. The method can be used to fabricate zone plates with more than 1000 zones, which cannot be done by other lithographic techniques [72]. The structure of the Gabor or holographic zone plate can be generated by recording the interference pattern produced by a plane or spherical wavefront (parabolic zone plates) or two spherical wavefronts (elliptical zone plates) propagating in opposite directions (figure 2.1). The Gabor amplitude zone plate with a sinusoidal absorption profile has a remarkable property: it produces no diffraction orders beyond the first (table 2.1). However, according to (2.8b), the theoretical diffraction efficiency of this plate is low, i.e. 6.25%, which is substantially less than the efficiency of binary zone plates. This applies to both amplitude and (particularly) phase plates ($\sim 40\%$).

Consider a Gabor phase plate that produces a continuous phase variation in accordance with (2.7b). The phase is varied by the varying optical path length in the layer, and this is described by (2.32) and (2.33). The overall amplitude transmission factor is complex and is given by the following expression:

$$\tilde{T} = \exp(-\chi\,\Delta\varphi + i\,\Delta\varphi). \tag{2.42}$$

The first term on the right represents absorption by the plate material.

Suppose that the radial variation in the thickness of the phase-shifting layer on the holographic zone plate is described by

$$t = t_0 + t_1 \cos(\pi r^2/\lambda F). \tag{2.43a}$$

The corresponding phase shift distribution is then given by

$$\varphi_1 = 2\pi\,\delta t_1/\lambda. \tag{2.43b}$$

Omitting the insignificant part of the constant phase shift, we can write

$$\tilde{T} = \exp\left[-\chi\varphi_1 \cos(\Omega r^2/F) + i\varphi_1 \cos(\Omega r^2/F) \exp(-\chi\varphi_1)\right] \tag{2.44}$$

where $\Omega = \pi/\lambda$.

Let us suppose that $\chi \ll 1$, which corresponds to a pure phase zone plate. In this case, $\tilde{T} = \exp[i\varphi_1 \cos(\Omega r^2/F)]$, so that expanding in the usual Fourier series, we have

$$\tilde{T} = \sum_{m=-\infty}^{m=\infty} i^m J_m(\varphi_1) \exp(im\Omega r^2/F) \tag{2.45}$$

where $J_m(\varphi_1)$ is the Bessel function of order m. For given m, the corresponding term in (2.45) is the amplitude in the mth-order diffraction. It follows from (2.45) that the diffracted radiation contains the higher diffraction orders and

$$I_m = I_0[J_m(\varphi_1)]^2. \tag{2.46}$$

The maximum first-order amplitude is reached for $\varphi_1 = 1.94$, so that the optimum thickness of the modulating layer is

$$t_{\text{opt}} = 1.94\lambda/2\pi\delta \approx 0.3\lambda/\delta. \tag{2.47}$$

For silicon and $\lambda = 0.834$ nm, the optimum modulation thickness is $t_{\text{opt}} = 1.44\,\mu$m.

This is smaller by a factor of 1.7 than is required for a binary phase zone plate, which simplifies the plate fabrication technology. However, the maximum possible theoretical diffraction efficiency of holographic phase zone plates is somewhat lower than for binary plates:

$$I_{1,\text{max}}/I_0 = 0.34. \tag{2.48}$$

2.3   KINOFORM ZONE PLATES

Our discussion of the zone plate has been based on the interference pattern produced in space by interfering plane and spherical waves (figure 2.1). This pattern provides us with a simple means of describing the structure of Fresnel zones in space. If we use parabolic and ellipsoidal coordinates, we can also describe the properties of 'kinoform' lenses based on the refractive properties of materials in the X-ray range. The basic principle underlying the kinoform lens is that the profile of the phase-shifting surface is arranged so that the transmitted waves combine at the focus with a constant phase (in the case of the zone plate, they combine with a constant phase sign). A jump in the phase is allowed only at points that are multiples of 2π (kinoform plate) or $2n\pi$ (Wood plate) where $n = 1, 2, 3$ etc.

This type of synthesized optics was used for the first time in [73] in which the word 'kinoform' was coined. A simple example of a kinoform focusing lens is shown in figure 2.14 in which sections of a thin ordinary lens of focal length F are projected onto a plane. The sections are made at right angles to the optical axis so that the phase difference between the wave fronts is 2π, which corresponds to a thickness of λ/δ ($m\lambda/\delta$ in the case of the Wood plate). The resulting stepped profile has thicknesses ranging from fractions of a micron to several microns, and consitutes a 'planar lens' with all the properties of the perfect thin parabolic lens. We shall now consider the properties of the kinoform

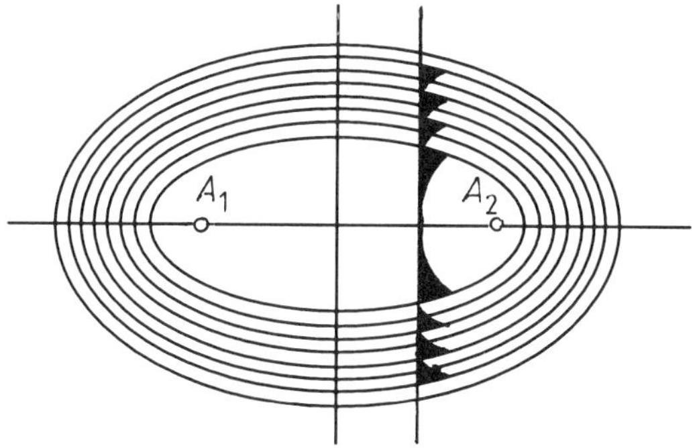

Figure 2.14   The principle of the kinoform lens.

lens as a focusing element. For the sake of simplicity, we will confine ourselves
to the axially symmetric case.

We first transform to parabolic coordinates:

$$r = \xi\eta \qquad y = (\eta^2 - \xi^2)/2. \tag{2.49}$$

If we suppose that the focal point is at the origin of coordinates, and the optical
element is located at a distance F along the y axis, the radial position coordinate
in the cross-sectional plane of a paraboloid is

$$r^2 = 2F\xi^2 + \xi^4 \tag{2.50}$$

whence

$$\xi_n^2 = -F + \sqrt{F^2 + r_n^2}. \tag{2.51}$$

This means that ξ^2 is the optical path difference and the phase shift is given by

$$\Delta\phi = 2\pi\xi^2/\lambda. \tag{2.52}$$

The condition for constant phase difference (to within 2π) in the focal plane
can be written

$$(2\pi/\lambda)(\sqrt{F^2 + r^2} - F) = 2\pi\delta t/\lambda \tag{2.53}$$

where t is the thickness of the phase-shifting material:

$$t = \left(\sqrt{F^2 + r^2} - F\right)/\delta. \tag{2.54}$$

The condition $\xi^2 = \lambda$ means that the phase shift is equal to 2π.

To estimate the maximum diffraction efficiency of the kinoform lens, we use
the formalism employed to investigate the properties of the binary phase zone

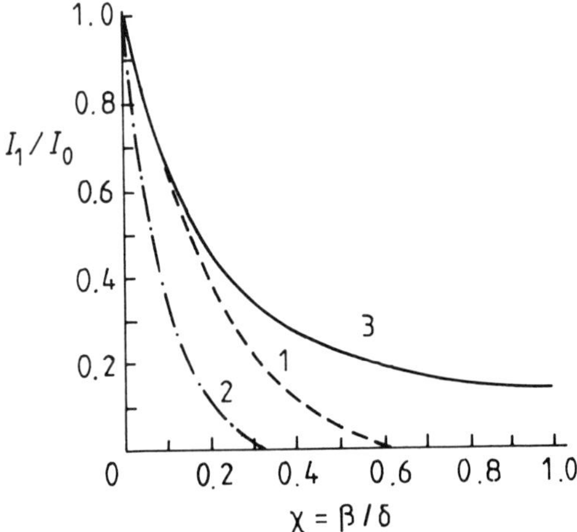

Figure 2.15 Efficiency of the kinoform zone plate (1), the Wood plate (2) and the optimized zone plate (3) as a function of χ.

plate. For a constant phase in the focal plane, the amplitude of the wave focused by the kinoform element is

$$E_1 = \frac{E_0}{2\pi} \int_0^{\sqrt{\lambda}} \exp(-2\chi\pi\xi^2/\lambda)\,\mathrm{d}\xi. \tag{2.55}$$

Figure 2.15 shows the calculated efficiency as a function of χ for the kinoform and Wood zone plates with a phase delay of 4π. The small difference between the efficiencies observed for small χ increases rapidly with χ, so that the kinoform lens is best fabricated from highly transparent materials. According to figure 2.15, kinoform plates are sufficiently efficient for $\chi \leqslant 0.15$.

The kinoform lens has several important advantages as compared with ordinary zone plates: (1) its aberrations are those of the perfect thin lens, (2) higher diffraction orders are absent, (3) the minimum size of a circular kinoform lens is greater by a factor of at least two as compared with the Fresnel zone plate, and by a factor of m as compared with the Wood kinoform plate. This result is very important for the technology used to fabricate high-resolution kinoform elements. The kinoform lens with the profile defined by (2.54) does not constitute the optimum design. The suggestion made in [74–76] is that part of the kinoform zone plate should be cleared by using composite interference. The principle of this lens is illustrated in figure 2.16. The dashed line shows the ideal profile of the kinoform lens with $\beta = 0$ (no absorption). The solid curve is the optimized profile with $\beta > 0$. Optimization was performed by varying the efficiency function. The final optimum thickness distribution of the plate was found to be [76]

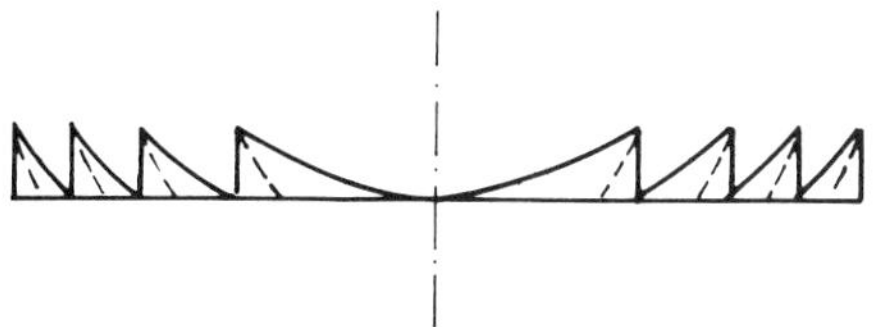

Figure 2.16 Profile of the kinoform lens for $\beta > 0$ (solid line) and the corresponding distribution for the optimized zone plate with $\beta = 0$ (dashed line).

Figure 2.17 Synthesized X-ray optical element.

$$t(r) = \begin{cases} 0 & r_1 - r_{\mathrm{opt}} \leqslant r \leqslant r_1 \\ (\sqrt{F^2 + r^2} - F)/\delta & 0 \leqslant r \leqslant r_1 - r_{\mathrm{opt}} \end{cases} \qquad (2.56)$$

where F is the focal length.

The optimum size of the transparent region, r_{opt}, can be determined from the following relations:

$$\exp[-2\pi\chi(L_{\mathrm{opt}} + 1)] = \sin(\gamma - 2\pi L_{\mathrm{opt}})/\sin\gamma \qquad (2.57)$$

$$\cos\gamma = \beta(\beta^2 + \delta^2)^{1/2} \qquad (2.58)$$

$$L_{\mathrm{opt}} = \sqrt{F^2 + (r_n - r_{\mathrm{opt}})^2} - \sqrt{F^2 + r_n^2}. \qquad (2.59)$$

For the sake of simplicity, we can confine our attention in (2.57)–(2.59) to the first kinoform zone with radius $r_1 = 2F\lambda + 2\lambda^2$ which corresponds to a phase shift of 2π in the diffracted wave. In the case of an absorbing phase zone plate, the ring of width r_{opt} is left unobstructed. The maximum thickness of the phase-shifting material can therefore be calculated from the formula

$$t_{\mathrm{max}} = \left[\sqrt{F^2 + (r_1 - r_{\mathrm{opt}})^2} - F \right] \Big/ \delta \qquad (2.60)$$

and depends on the quantity r_{opt}. According to [76], the efficiency of the optimized kinoform element is

$$I_1/I_0 = (1/\pi^2) \sin^4(\pi L_{\mathrm{opt}})(1 + 1/\chi^2). \qquad (2.61)$$

Figure (2.15) also shows the efficiency of the zone plate as a function of the parameter χ of the material. It follows from this that there is practically no need for optimization when $\chi = 0.1$, and the efficiency reaches 65%.

The kinoform optical elements described above constitute a simple example of the amplitude-phase surface profile being used for the transformation of wavefronts. The so-called 'computer optics' have recently gone through major advances in relation to the transformation of complex waveforms [77, 78]. Synthesized amplitude-phase optical elements can be very useful in 'he X-ray range, e.g. as a means of producing a given intensity distribution in the focal plane [79]. There are many situations, e.g. in the case of the X-ray microprobe, in which an optical element is not required to produce an image. The main requirement is then the production of an ideal spherical wavefront with a focal spot of minimum size. Computer optics can then be useful, e.g. in transforming the complex shape of the X-ray source into a converging spherical wavefront.

2.4   ZONE PLATES WITH A LINEAR FOCUS

There is considerable practical interest in 'linear' zone plates for which the energy distribution in the focal plane takes the form of a segment of a line. Let us suppose that the source of radiation is an infinite line passing through the point A_1, at right angles to the optical axis A_1–A_2 (figure 2.12(a)). Conventional X-ray tubes constitute a complex source shape. We shall suppose that the image of the source is also a line crossing the axis at A_2. The surfaces of equal phase are then cylinders with elliptical cross sections in the plane perpendicular to the source line. The linear zone plate is formed by cutting the surfaces of equal phase with a plane perpendicular to the optical axis, as in the case of the concentric zone plate. The structure produces a line focus. We shall now examine in detail a useful feature of the Fresnel zone plate [80, 81].

If the linear plate is placed at an angle α to the beam axis (figure 2.18), the result is a much higher resolution for the same zone dimensions. For the same focal length and minimum zone dimensions, the increase in the effective lens aperture is described by

$$A(\alpha) = A_0/\sin\alpha \qquad (2.62)$$

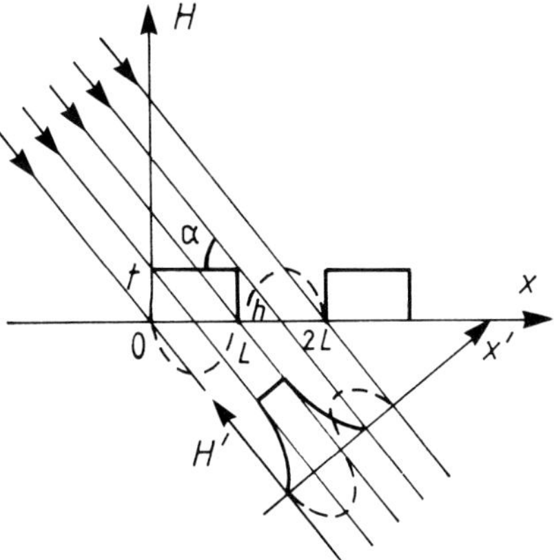

Figure 2.18   The contrast of the linear zone plate in the case of inclined incidence.

where A_0 is the aperture of the zone plate perpendicular to the optical axis with the same parameters δr_n and F. A resolution of, for instance, 50 nm can be achieved by using small grazing angles and high-resolution lithography. Let us estimate the maximum resolution and efficiency of the linear zone plate inclined to the beam axis. For a gold line of thickness t and minimum zone width δr_n, the resolution of the linear zone plate is determined by the inclination of the plate to the axis, i.e. by the projection of the zone-plate element onto the direction perpendicular to the beam axis. The effective zone size is thus found to be $\delta r'_n = \delta r_n \sin \alpha$. Hence, by reducing the grazing angle we can significantly increase the resolution of the linear zone plate. However, it is important to take account of the geometrical shadow cast by a grating line, which reduces the efficiency (figure 2.18).

Since the intensity of radiation transmitted by a zone covered by a masking layer with a linear absorption coefficient μ_t is $I = I_0 \exp(-\mu_t t)$, the intensity in the focal plane is

$$I(\alpha) = (\delta r_n - t \cot \alpha)[1 - \exp(-\mu_t t / \sin(\alpha)] \qquad (2.63)$$

where δr_n is the width of a line on the grating. The relative loss of intensity as compared with the case of normal incidence on the grating is given by

$$\frac{I(\alpha)}{I(0)} = \left(1 - \frac{t}{\delta r_n} \cot \alpha\right) \frac{1 - \exp(-\mu_t t / \sin \alpha)}{1 - \exp(-\mu_t t)}. \qquad (2.64)$$

Because of the greater effective aperture of the zone plate inclined at an angle α, it turns out that for equal focal lengths, equal thicknesses of the masking layer on the obstructed zones and maximum technological resolution, the efficiency factor of an inclined linear zone plate is

$$\frac{I(\alpha)}{I(0)} = \frac{1}{\sin \alpha} \left(1 - \frac{t}{\delta r_n} \cot \alpha\right) \frac{1 - \exp(-\mu_t t / \sin \alpha)}{1 - \exp(-\mu_t t)}. \qquad (2.65)$$

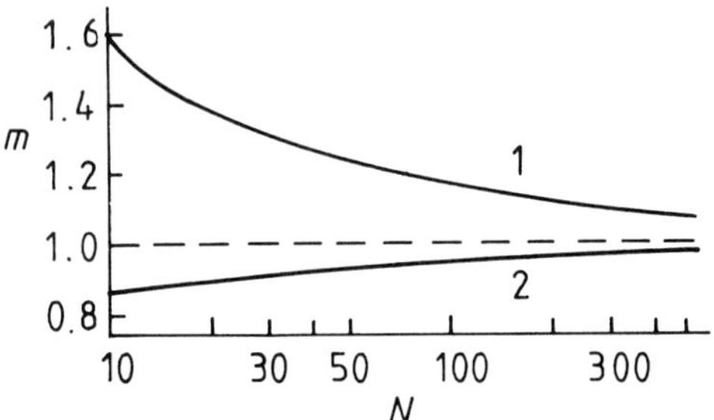

Figure 2.19   Transparency of linear zone plates as a function of the number N of zones: (1) positive plate, (2) negative plate.

For example, for $\alpha = 30°$ and $t/L = 0.1$, we have

$$I(\alpha)/I(0) \approx 1.3[1 - \exp(-\mu_t t)] \tag{2.66}$$

i.e. there is practically no loss of intensity when the resolution rises by a factor of two.

The overall intensity of the linear zone plate (LZP) with an absorbing layer is different from the efficiency of a concentric zone plate (CZP). For the latter, the areas of all the zones are equal, so that 50% of the incident radiation is lost. In the case of the linear zone plate, the loss of radiation in negative (obstructed) and positive (first zone unobstructed) plates is very different. The areas of the odd and even zones are respectively

$$S^{(1)} = L_x\sqrt{F\lambda}(\sqrt{2n+1} - \sqrt{2n}) \tag{2.67}$$

$$S^{(2)} = L_x\sqrt{F\lambda}(\sqrt{2n+2} - \sqrt{2n+1}) \tag{2.68}$$

where L_x is the length of the linear zone plate. The transparency factor m of the LZP and CZP is defined by

$$(I_1/I_0)_{\text{LZP}} = m(I_1/I_0)_{\text{CZP}}$$

where

$$m_{\text{neg}} = 0.5(1 + S^{(2)}/S^{(1)}) \qquad m_{\text{pos}} = 0.5(S^{(1)}/S^{(2)})(1 + S^{(1)}/S^{(2)}).$$

Figure 2.19 shows m_{neg} and m_{pos} as functions of the zone number. These values may be used to convert the efficiencies listed in table 2.1. The efficiency remains the same for pure phase and kinoform LZP.

3

X-ray image transfer by holography and coherent optics

In this chapter we consider applications of holography and coherent optics to the formation and transfer of images in the X-ray range. We present experimental results drawn from X-ray holography, and also from the Fourier images of periodic structures. Experimental approaches to the evaluation of image-structure parameters are discussed. The influence of the coherence properties of radiation emitted by a real source on image transfer is analysed. Techniques for producing non-coherent Fourier images and their experimental verification are presented. Different applications of coherent X-ray optics are discussed.

The holographic image differs from the image produced by contact or scanning microscopy in that it contains information about the spatial structure of the object. The hologram also provides information about the phase distribution in the object that cannot be obtained by conventional methods. The principles of holography, developed in the 60s and 70s, have led to the creation of a new branch of optics that is actively concerned with the development of coherent optical systems for data transfer, processing and storage [82].

3.1 METHODS OF X-RAY HOLOGRAPHY

The potentialities of X-ray holography and interference microscopy can be conveniently examined by dividing the X-ray range into two ranges, namely, hard X-rays ($\lambda < 0.1\,\text{nm}$) and soft X-rays ($\lambda \approx 0.1\text{--}20\,\text{nm}$).

Comparisons of recent experimental studies with theoretical calculations have shown that conventional optical holography is unlikely to be successful in the hard X-ray range. Earlier work demonstrated that the resolution of traditional systems in the X-ray range did not exceed $1\text{--}10\,\mu\text{m}$ because existing X-ray sources did not produce suffient X-ray power. The low absorption coeefficient and the magnitude of the refractive index at wavelengths below $0.3\,\text{nm}$ ensure that an increase in resolution by only an order of magnitude requires an increase in X-ray output power by a factor of 10^6 [83].

3.1.1 X-ray holograms

There are no optical elements, in the usual meaning of the phrase, that can be used in the hard X-ray range ($\lambda < 0.1\,\text{nm}$). Reflection, splitting into two coherent beams and even a degree of focusing in the X-ray range can only be achieved with the aid of perfect, defect-free crystals. Total external reflection by the surface of a metal has found very limited application as the basis for specularly reflecting elements. The angle of reflection in such cases is no more than a few dozen minutes of arc. Diffraction focusing by perfect crystals, on the other hand, is being intensively developed as the basis for X-ray optical elements. The first successful experimental attempt to focus a spherical wave in perfect silicon crystals was reported in 1978 [84]. The width of the focus was reduced down to $10\,\mu\text{m}$. Subsequent experiments with monochromatic radiation ($\Delta\lambda/\lambda = 10^{-5}$) reduced this to $3\,\mu\text{m}$.

The position of the focused image can be varied by varying the thickness of the crystal or by bending it. The diffraction efficiency of an X-ray optical element can reach 25%, but it must be emphasized that the diffraction optical element is not an analogue of a lens, since it can only actually produce a wave with inverted curvature. A plane wave is transformed into a plane wave. Coherent optical systems for the X-ray range can only be based on diffraction by perfect crystals. The high-resolution interference X-ray microscope is an example of this type of device (figure 3.1). Polychromatic radiation from a point source is diffracted by a thin crystal C_1 and secondary diffraction is produced by 'mirrors' M. Radiation of each wavelength is thus compressed into a small spot by diffraction focusing. The spot becomes the source of a spherical wave, the source width being of the order of $10\,\mu\text{m}$. The object is inserted into the spherical wave and a magnified image of the object with the coherent field F superimposed upon it is produced after the crystal C_2. This arrangement is the simplest holographic microscope using a hologram of a magnified focused image. The image-magnifying system is employed in order to reduce aberrations introduced by diffraction by the different crystals in the interferometer.

Most X-ray holographic experiments have been performed in the wavelength range $\lambda = 1\text{–}10\,\text{nm}$. This range has attracted considerable attention for the following reasons:

(1) The absorption coefficient of organic and inorganic materials for soft X-rays is relatively high so that high-contrast images can be produced.
(2) Organic, including biological, objects are less damaged by X-rays than by electrons in the electron microscope.
(3) There are sufficiently efficient optical elements for the focusing and large-angle reflection of soft X-rays.

The following systems have been used for the holographic recording of X-ray images:

(a) *Lensless Fourier holography (figure 3.2(a))*
The spatial resolution of this system is independent of the resolution of the

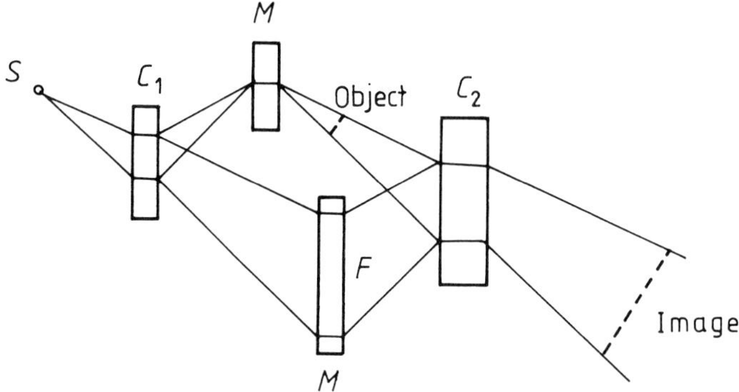

Figure 3.1   Principle of the X-ray interference microscope.

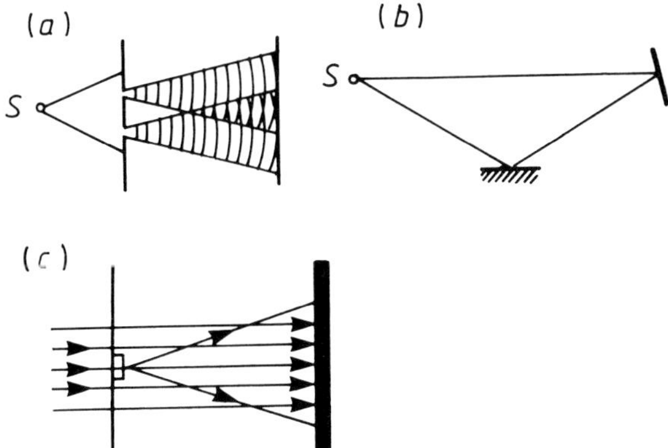

Figure 3.2   Methods used to produce X-ray holograms: (a) lensless Fourier holography [85], (b) Lloyd's mirror [86], (c) axial Fraunhoffer holography [87].

recording device and is wholly determined by the size of the reference 'point' source and the coherence of the radiation employed. The experimental resolution is 2.8 μm [85].

(b) *Hologram recorded with Lloyd's mirror (figure 3.2(b))*
The resolution of this system is limited by the angle of total external reflection and the coherence of the radiation. It has not exceeded 1 μm [86].

(c) *Axial Fraunhoffer holography (figure 3.2(c))*
Here the resolution is determined mostly by the resolution of the recording device (6 μm) and the coherence of the X-rays [87].

Good, high-resolution images are produced by *a posteriori* processing of X-ray images by coherent optical methods [88].

In all these experiments, the holographic image is reconstructed in visible light (whose wavelength is greater by two orders of magnitude than the X-ray

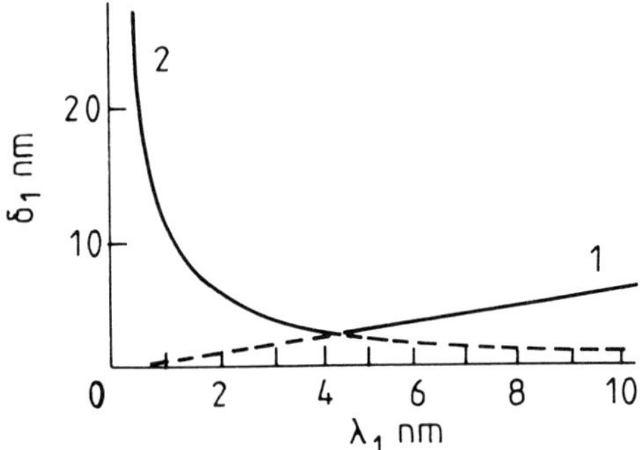

Figure 3.3 Resolution of the PMMA photoresist as a function of wavelength: (1) diffraction limit of resolution, (2) limit of resolution restricted by the free path range of secondary electrons in the resist.

light (whose wavelength is greater by two orders of magnitude than the X-ray wavelength), which gives rise to unavoidable aberrations in the reconstructed image [89].

The main factors that restrict the resolution of X-ray holographic methods are therefore the following:

- the resolving power of the recording system,
- the method used for holographic image reconstruction,
- coherence of X-ray sources.

Further improvements in the resolution of the methods used in X-ray holography will depend on these factors.

We shall now examine the characteristics of the elements of a hologram writing system by considering the example of axial Fraunhoffer holography. This system is the least sensitive to the coherence of the X-ray radiation. Its resolution is determined by the following parameters:

$$\delta_{x1} = 1.22\delta_r R_1/(R_1 + R_2) \tag{3.1}$$

$$\delta_{x2} = 1.22d_s R_2/(R_1 + R_2) \tag{3.2}$$

$$\delta_{x3} = 0.43\sqrt{R_2 \Delta\lambda} \tag{3.3}$$

where R_1 is the source–object separation, R_2 is the object–recorder separation, $\delta_{x1}, \delta_{x2}, \delta_{x3}$ are, respectively, the resolutions due to the registrator, the spatial coherence and the temporal coherence of the radiation, δ_r is the resolution of the registrator in lines per mm, d_s is the size of the source and $\Delta\lambda$ is the line width of the radiation.

The longitudinal resolution of the resonstructed image is given by

$$\delta_z = (R_2) \Delta\lambda/4\lambda. \tag{3.4}$$

Nuclear photographic emulsions have a resolution of up to 250 lines per mm. In the case of the X-ray photoresist, the resolution is of the order of 2×10^5 lines per mm. It is therefore useful to consider the possibility of using X-ray photoresists for writing X-ray holograms.

Figure 3.3 illustrates the limiting resolution in the interference pattern written on the PMMA X-ray photoresist. Curve 1 shows the pure diffraction limit of resolution, set by the wavelength, and curve 2 shows the limit of resolution, set by the range r_{se} of secondary electrons in the resist, which is given by

$$r_{se} = 3.43 \times 10^3/\rho\lambda^{1.35} \tag{3.5}$$

where ρ is the density of the photoresist in $\mathrm{g\,cm^{-3}}$.

The data of figure 3.3 shows that the region of maximum resolution occurs in the wavelength range 3–6 nm. The resolution is then 5 nm. Photoresists are difficult to work with in practice because of their low sensitivity.

Let us now estimate the X-ray source brightness that is necessary for given resolution:

$$B_\lambda = D_0 R_1^2 \lambda R_2/4\mu_t d_s^2 \tau \delta_x^2 \tag{3.6}$$

where δ_x is the required transverse resolution, τ is the exposure time, D_0 is the dose sensitivity of the photoresist and μ_t is the linear absorption coefficient. If we suppose that the resolution is 50 nm, the wavelength is 4 nm, the object–hologram separation is 0.1 mm and the photoresist is PMMA ($D_0 = 100\,\mathrm{J\,cm^{-3}}$), we find that the necessary brightness is $B_\lambda = 10^8\,\mathrm{W\,cm^{-2}}$.

(d) *Reconstruction of the X-ray hologram*

The interference pattern (hologram) is written on the resist in the form of a surface profile. The subsequent reconstruction and magnification of the object recorded on the hologram can be performed by the method developed for contact X-ray microscopy [90]. The surface of the developed resist is coated with a thin film of gold and is examined in a scanning electron microscope [91]. The image reconstruction procedure can also be implemented on a computer, using the analogue signal from the SEM. Figure 3.4(a) shows a fragment of the surface of a metal-coated resist with the hologram of a pinhole written upon it. The image reconstructed from this hologram is shown in figure 3.3(b) [92]. The size of the pinhole is $10\,\mu\mathrm{m}$, so that these photographs only demonstrate the feasibility of this method of reconstruction. It is important to note that the possibilities of optical reconstruction are limited, so that it is best to use numerical reconstruction on a computer.

In optical reconstruction of holograms written by the Fraunhoffer method we encounter the problem of separating superimposed real and virtual images. This difficulty can be obviated by using counter-propagating beams and an inclined reference beam to write the hologram (figure 3.5) [93]. As the X-ray beam crosses the resist, it is diffracted by the object. The interference mirror reflects the diffracted radiation, which constitutes the object wave. The interference

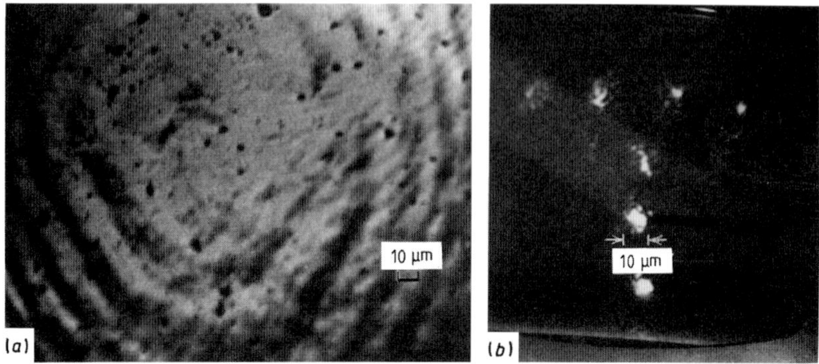

Figure 3.4   The surface of a metallized resist bearing the hologram of a pinhole (*a*) and the image reconstructed from the hologram (*b*) [92].

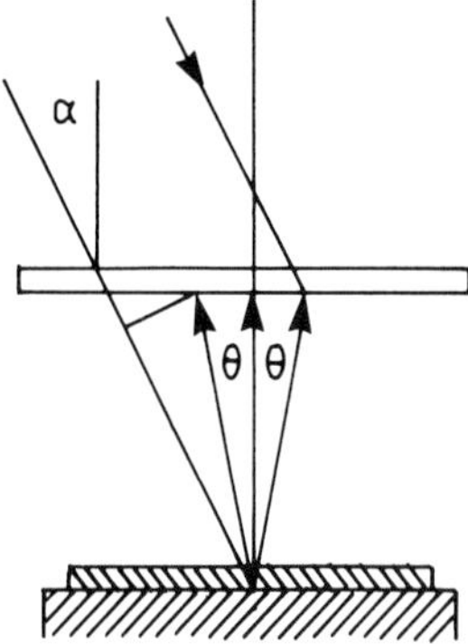

Figure 3.5   Hologram written with counter-propagating beams [93]:   (1) multilayer interference mirror, (2) object, (3) resist.

pattern is then processed by the method described above and is reconstructed by the conjugate reference wave.

3.1.2   Sources of coherent X-rays

Coherent X-ray sources, i.e., lasers, are not as yet available although much effort is being devoted to the development of such sources.   Non-coherent conventional sources are therefore commonly used in X-ray microscopy and holography, among which X-ray tubes produce the most highly monochromatic radiation and have smallest radiating element.

(a) *X-ray tubes*
These rely on the emission of characteristic X-ray lines by the anode material. The lines are quite sharp (usually $\Delta\lambda/\lambda = 0.5 \times 10^{-3}$) and the strongest of them are emitted in transitions between the atomic L and K shells (Kα lines). These characteristic lines are superimposed on the broad bremsstrahlung continuum

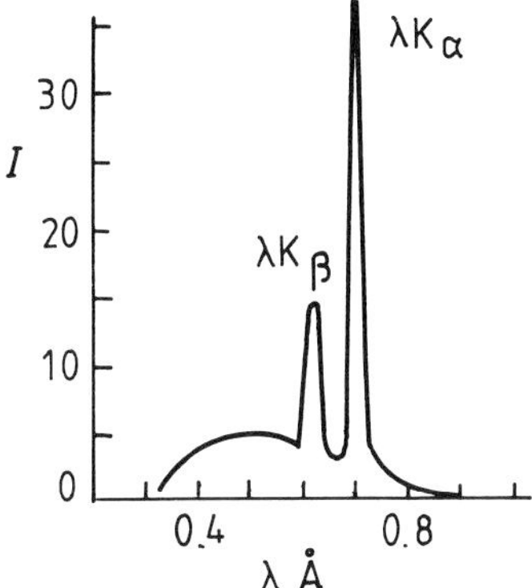

Figure 3.6 The spectrum produced by an X-ray tube with a molybdenum anode.

emitted by the X-ray tube (figure 3.6). In the absence of a filter, the natural ratio of the peak intensity of a characteristic X-ray line to the bremsstrahlung intensity is of the order of 100. The tube emission efficiency can be calculated from the empirical formula

$$n_{\text{phot}}/n_{\text{e}} = 3 \times 10^{-4}(W - W_K)^{1.63} \exp(-0.095Z)$$

where W is the energy of the electron beam, W_K is the energy of the K-level participating in the transition and Z is the atomic number of the anode material. The disadvantage of the X-ray tube is that its radiation oputput is limited because the anode is heated by the incident electron beam. Moreover, the emitted radiation is not directional, so that only a relatively small fraction of this radiation falls into any given small solid angle. The heating problem can be solved to some extent by using a rotating anode. For an angular velocity of up to 1000 rpm, the output power can be raised in this way by two orders of magnitude. The resulting spectral flux is 3×10^{10} photons per second [94].

(b) *Synchrotron radiation*
Synchrotron radiation, i.e., the radiation emitted by relativistic electrons in a magnetic field, has a high intensity. It is generated because the relativistic electrons are accelerated radially by the bending magnet [95]. When a high enough vacuum is employed ($\sim 10^{-9}$ Torr), the electron beam persists with constant energy for a period of the order of 10 h. A storage ring can be regarded as a quasi-continuous source of radiation (it must, however, be remembered that the time structure of bunches is very important in some experiments). The synchrotron radiation emerging from the bending magnet has a relatively wide spectrum (figure 3.7), but is compressed in the vertical direction in which it is confined to an angle $\theta_{\text{e}} = 1/K$ where the rigidity parameter is $K = 1957\,W$. For example, when $W = 1\,\text{GeV}$, we find that $\theta \approx 5 \times 10^{-4}$. Crystals or diffraction gratings are used as monochromators for synchrotron radiation, but the use of such additional elements results in considerable loss of intensity. Moreover, the

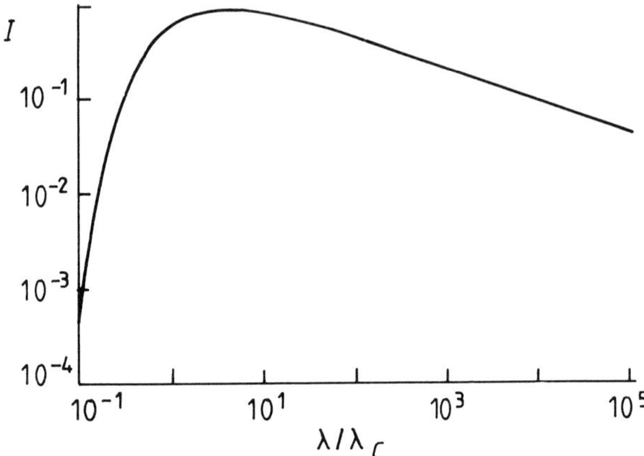

Figure 3.7 The spectrum emerging from the bending magnet of a synchrotron source [96].

effective size of the emitting region in the storage ring is 1–2 mm, which is much greater than the size of the radiating region of an X-ray tube. This means that the working distance between source and hologram must be increased to ensure an acceptable range of coherence. Despite all this, the spectral intensity is higher by approximately an order of magnitude as compared with the microfocus X-ray tube, demostrating the considerable advantages of synchrotron radiation in X-ray holography [96, 97].

In the most powerful source of X-rays available at present, synchrotron radiation is produced by the so-called undulator [98], which relies on a periodic modulation of the magnetic field along the path of an electron beam. When this variation in the magnetic field is nearly sinusoidal, the emission spectrum takes the form of a narrow line ($\Delta\lambda/\lambda$ equal to the number of periods) of wavelength given by

$$\lambda = \Lambda_0/2K^2 \tag{3.7}$$

where Λ_0 is the modulation period. The emitted radiation is highly directional and is confined to a solid angle $\Omega_0 \approx \pi/K^2$. By suitably choosing the magnetic field, it is possible to produce horizontal, vertical or circular polarization. As an example, we list the parameters of the undulator mounted in the beam line of the photon factory at Tsukuba in Japan [117]:

Total length	3.8 m
Period	6 cm
Number of periods	60 (number of poles 119)
Magnetic material	$SmCo_5$
Magnet gap range	2.7–8 cm
Range of values of K	1.78–0.1

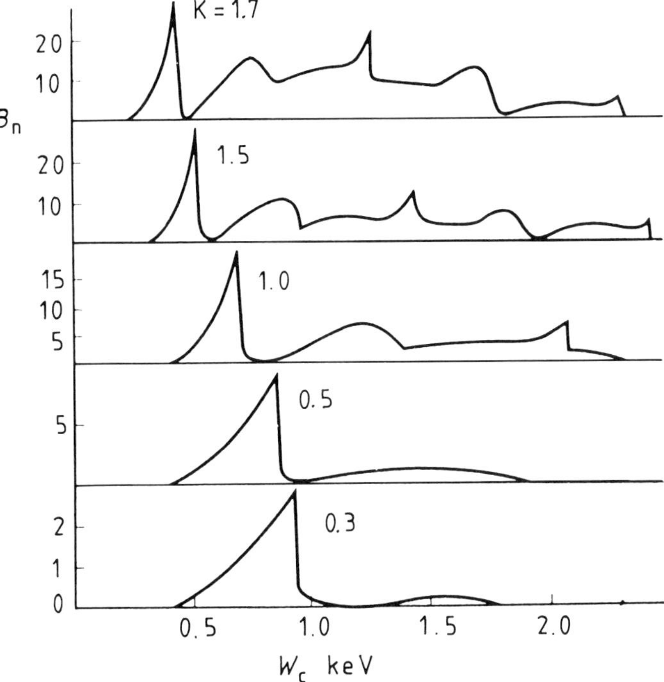

Figure 3.8   The spectrum of radiation from an undulator for different values of the rigidity K.

The spectrum produced by an undulator consists of a number of peaks with intensities given by

$$B_n = 1.74 \times 10^{12} W^2 H_0 i \, \Phi_n(K)$$

where W is the energy of the electrons in GeV, i is the electron current in mA, K is the undulator rigidity, H_0 is the magnetic field intensity in the magnet gaps in tesla and B_n is measured in photons per second per millisteradian in a band of $\Delta\lambda/\lambda = 0.01$ and beam current of $1\,\text{mA}$. The rigidity K determines the spectrum of the radiation from the undulator in terms of the influence function

$$\Phi_n(K) = \frac{n^2 K^2}{1 + K^2/2} \left\{ J_{(n-1)/2}\left[\frac{nK^2}{1 + K^2/2} \right] - J_{(n+1)/2}\left[\frac{nK^2/4}{1 + K^2/2} \right] \right\}.$$

Figure 3.8 shows the X-ray spectrum from an undulator for different values of the rigidity K in the range 0.3–1.7. The measured radiation wavelength is shown in figure 3.9 as a function of the undulator magnet gap for three harmonics ($\lambda_{1,2,3}$ are the three leading harmonics). The first harmonic has the highest intensity. The mean spectral width of the first harmonic is $\Delta\lambda/\lambda = 0.11$ [117].

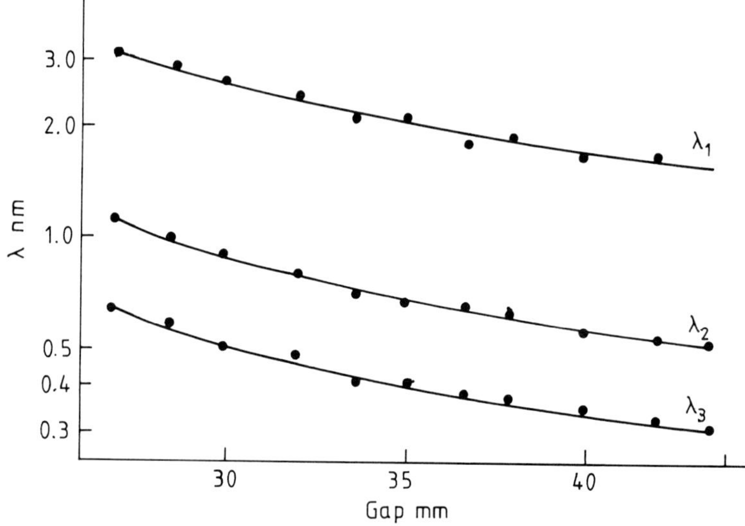

Figure 3.9 Radiation wavelength as a function of magnet gap size for the three leading harmonics of an undulator [117].

Finally, we need to say something about pulsed X-ray sources. A very interesting source of this kind employs the radiation emitted by plasma heated to a high temperature by an electric discharge in a capillary. The discharge is initiated by an electron beam or by radiation from a powerful laser [99]. Such sources produce the highest available spectral radiation intensity. Comparison of actual X-ray source intensities with estimates based on Fraunhoffer holograms written with a resolution of 50 nm shows that a holographic experiment with this order of resolution is possible only if we use undulator radiation.

3.2 APPLICATION OF THE TALBOT EFFECT TO X-RAY IMAGE TRANSFER

One of the coherent-optical X-ray phenomena that has attracted considerable attention is the diffractive reconstruction of the images of periodic structures. This can be described as a method of producing holograms without a reference beam, in which the reconstruction of the object relies on some of the radiation from the object itself.

3.2.1 Fourier imaging in coherent radiation

The Fourier image effect (Talbot effect) has recently been discussed by many reseachers in connection with possible applications to image transfer at short wavelengths [100–109]. The essence of this effect is that, when a periodic object is illuminated by a plane monochromatic wave, its image observed without the assistance of any optical elements is exactly repeated at distances given by the formula

$$R_{n,m} = \frac{md_M^2}{n\lambda} \qquad (3.8)$$

where d_M is the period of the object, λ is the wavelength and $n = 1, 2, 3, ...$

The effect was first explained by Rayleigh as far back as 1881 and the theory of it was subsequently developed further by Cowley [100]. The Talbot effect can be described in terms of Fresnel diffraction, using the small-angle approximation in the Kirchhoff integral. For the sake of simplicity, we consider the one-dimensional case. If x represents the plane of the object and u the the plane of the image, the amplitude $E(u)$ in the image plane is related to the object transmission function $g(x)$ as follows [100]:

$$E(u) = \left[\frac{i}{R\lambda}\right]^{1/2} \exp\left[-i\frac{2\pi R}{\lambda}\right] \int_{-\infty}^{+\infty} g(x) \exp\left[-i\frac{\pi(u-x)^2}{R\lambda}\right] dx. \qquad (3.9)$$

When the object transmission function is $g(x) = \cos(2\pi hx/d_M)$, we can readily show that the intensity in any image plane is given by

$$I(u) = \cos^2(2\pi u/d_M). \qquad (3.10)$$

However, the transmission function of a periodic object can be expanded into a cosine series:

$$g(x) = \sum_{h=0}^{\infty} F_h \cos(2\pi hx/d_M) \qquad (3.11)$$

so that, by evaluating the integral in (3.9), and using (3.11), we obtain

$$E(u) = \exp\left[-i\frac{2\pi R}{\lambda}\right] \sum_{h=0}^{h=\infty} F_h \exp\left[i\frac{\pi\lambda h^2}{d_M^2}\right] \cos\left[\frac{2\pi hu}{d_M}\right]. \qquad (3.12)$$

Hence it is clear that if $R\lambda/d_M^2$ is an integer, the intensity distribution repeats the object transmission function. When $R = nd_M^2/m\lambda$, the period of the image is reduced by the factor m.

Let us now consider the integral in (3.9):

$$\int_{-\infty}^{+\infty} g(x) \exp\left[-i\frac{\pi(u-x)^2}{R\lambda}\right] dx.$$

This is the Fourier transform of the expression

$$g(x)\left(\frac{-i\pi x^2}{R\lambda}\right) \qquad (3.13)$$

multiplied by a certain phase factor.

We have already shown that for certain definite $R_{n,m}$ and infinite limits of integration, we have

$$g(u) = F\left[g(x)\exp\left(-i\frac{\pi x^2}{R\lambda}\right)\right]. \tag{3.14}$$

However, in practice, the object dimensions are always finite, so that $g(x)$ is a certain function $g'(x)$ multiplied by $\text{rect}(x/A)$ where A is the size of the mask ($A = md_M$) [110]. In accordance with the well known properties of the Fourier transformation, we can now write $E(u)$ as the convolution of Fourier transforms:

$$E(u) = g(u) * A\,\text{sinc}[\pi Au/R\lambda \tag{3.15}$$

where

$$A\,\text{sinc}\left[\pi Au/R\lambda\right] = \frac{\sin(\pi Au/R\lambda)}{\pi u/R\lambda}.$$

Accordingly, the function $A\,\text{sinc}[\pi Au/(R\lambda)]$ is the transfer function of the system and determines the modulation transfer frequency function of the process used to produce the Fourier images, i.e.,

$$H(v) = \begin{cases} 1 & v < A/(2R\lambda) \\ 0.5 & v = A/(2R\lambda) \\ 0 & v > A/(2R\lambda) \end{cases} \tag{3.16}$$

where $H(v)$ is a measure of the resolving power of the image transfer process. It follows from this expression for $H(v)$ that the best resolution is obtained in the first self-reproduction plane $n = 1$, whereas on intermediate image planes the resolution is increased by a factor of m and the period is reduced by the same factor (3.8).

The image defocusing effect can readily be estimated. As for the perfect thin lens, the defocused image can be described by the convolution

$$E(u) = E_0(u) * \exp\left(-i\frac{\pi u^2}{\lambda\Delta R}\right). \tag{3.17}$$

The fall in resolution due to the change ΔR is $\delta_z = \sqrt{\lambda\Delta R}$. We note that the small-angle approximation is valid so long as

$$\arcsin(\lambda/\delta_x) < \lambda/\delta_x < \delta_x/2R.$$

Hence, it follows that $\delta_x \approx 10\lambda$ is the theoretical limit of resolution in this method of image transfer.

If R_1 is the separation between the mask and the source of radiation (source size b_s), the number of structure periods illuminated coherently is given by

$$N \approx \lambda R_1 / (d_M / b_s) . \tag{3.18}$$

The maximum spatial resolution in the image is then given by

$$\delta_x = b_s d_M^2 / (\lambda R_1) . \tag{3.19}$$

When $\lambda = 10\,\text{nm}$, $R_1 = 10\,\text{cm}$, $b_s = 100\,\mu\text{m}$, $d_M = 1\,\mu\text{m}$, we find that the resolution is $\delta_x = 0.1\,\mu\text{m}$.

We have used model experiments with an optical source of coherent radiation (a laser) and a mathematical simulation of diffraction by one-dimensional and two-dimensional structures to determine the limits of projection lithography of periodic structures [111].

The mathematical simulation was performed on a computer, using a standard fast Fourier transform algorithm. For a given object transmission function

$$g(x) = \sum_{n=1}^{N} \text{rect}[4(x/d_M) - n] \tag{3.20}$$

we find that the intensity distribution in the image plane is

$$|E(u)|^2 = \left| \left[\frac{i}{R\lambda} \right]^{1/2} \int_{-\infty}^{+\infty} g(x) \exp\left[-i\frac{\pi x^2}{R\lambda} \right] \exp\left[i\frac{2\pi ux}{R\lambda} \right] dx \right|^2 . \tag{3.21}$$

We assumed in the calculations that the step size in x in the object plane was equal to the step size in $\omega = 2\pi u/R\lambda$ in the image plane. The two step sizes were related in the fast Fourier transform algorithm as follows:

$$\Delta\omega = 2\pi / (h\Delta x)$$

where h is the number of points at which the object function was specified.

It is clear from the expression for $E(u)$ that

$$\Delta\omega = \frac{2\pi}{\lambda R} \Delta u \tag{3.22}$$

i.e., $\Delta u = \lambda R/(h\Delta x)$.

The plane of the image is the first self-reproduction plane ($R_2 = d_M^2/\lambda$). Since $\Delta u = \Delta x$, we obtain the object period as a function of the number of points at which the object function is specified: $d_M = h^{1/2}$. It was assumed in the calculation that $h = 1024$ and $d_M = 32$.

Figure 3.10 shows the calculated image for different values of the number N of periods in the original structure. The number of periods in the plane of

Table 3.1

Source	λ nm	$\Delta\lambda$ nm	δ_x nm*	ΔR nm
Al Kα	0.83	0.006	50	2.99
C Kα	4.48	0.0300	80	1.43
Be Kα	11.40	0.7000	250	5.48
Synchrotron	4.50	0.0030	25	0.13

* $d_M = 1\,\mu$m.

the image is greater by one because the image is shifted by $\pi/2$ in the first self-reproduction plane, but remains symmetric relative to the object axis.

Figure 3.11 shows an example of image transfer for different values of the number N of mask periods. Figure 3.12 shows the limiting resolution of the method as a function of the number of mask periods. This is described by

$$\delta_x \approx d_M \left[\Delta\lambda/(2\lambda N)\right]^{1/2}. \tag{3.23}$$

Table 3.1 shows the calculated resolution parameters of the method for different X-ray sources ($\Delta R = \delta_x^2/\lambda$ is the longitudinal resolution).

The Talbot effect was successfully used in [105] to produce a one-dimensional periodic structure with a period of 20 nm in vacuum ultraviolet. However, the coherence of the source used in [105] was not good enough to produce a two-dimensional Fourier image.

The first successful experiments on the production of two-dimensional Fourier images were performed with the Rigaku soft X-ray microfocus source. The experimental arrangement is illustrated in figure 3.13. The electron beam from the lanthanum hexaboride electron gun was focused on an aluminum foil in a spot 5 μm in diameter. The accelerating voltage was 15 kV, the beam current was 400 μA and the thickness of the aluminium anode membrane was 7μm. The object was a copper grid, 4.7μm thick and was produced by electrolytic deposition (figure 3.14). Because of the considerable thickness of the anode membrane, the size of the X-ray source was estimated as 10 μm. To achieve a satisfactory coherence area, the recorder was at a distance of 872 nm from the source (this distance was limited by the overall length of the vacuum chamber).

The object–source and the object–recorder distances R_1 and R_2 were calculated from the following formula for a Fourier image in a divergent beam [112]:

$$\frac{1}{R_1} + \frac{1}{R_2} = \frac{1}{F_M} \tag{3.24}$$

where $F_M = d_M^2/m\lambda$ is the effective focal length of the mth-order Fourier image. For a given source–detector distance ($R_1 + R_2 = 872$ nm) and wavelength 0.834 nm (Kα line of aluminum), we have $F_1 = 193.4$ mn, $F_2 = 96.7$ nm. The image magnification is $(R_1 + R_2)/R_1$. Accordingly, we find that for the first

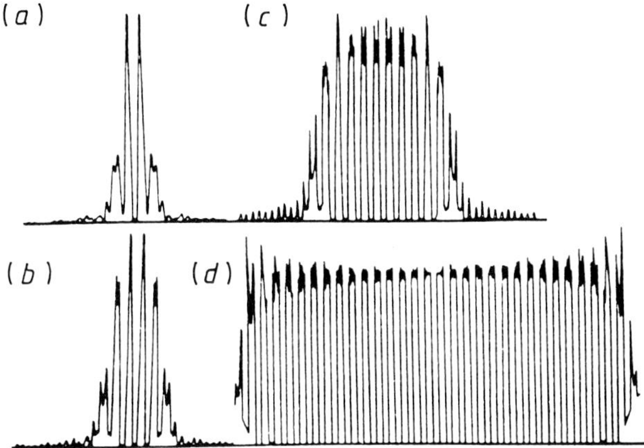

Figure 3.10   Calculated Fourier images of a periodic structure with the following values of the number of periods: $N = 1(a)$, $3(b)$, $9(c)$ and $30(d)$.

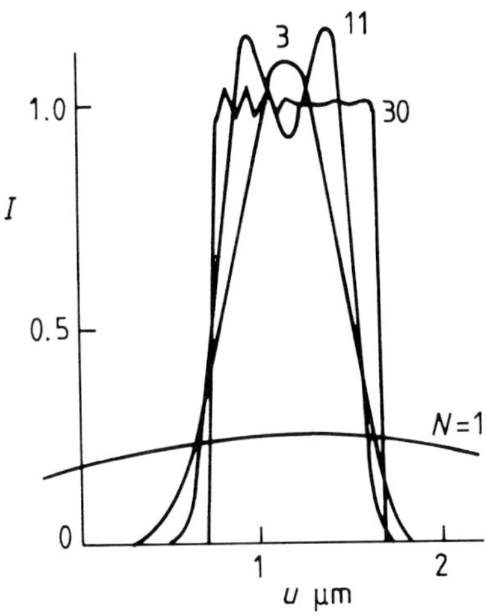

Figure 3.11   The shape of the central element of a periodic structure for different values of the number N of mask periods.

Fourier image, the source–object separation should be $R_1 = 582.5$ nm whereas for the second order $R_1 = 761.2$ nm. Using the estimated size of the source, the beam coherence radius in the plane of the detector is found to be

$$r_{coh} = (\lambda/b_s)\,(R_1 + R_2) \approx 72\,\mu m.$$

This value of R_{coh} is sufficient for a resolution $\delta_x = 2.5\,\mu m$ (according to (3.19)). The detector used in this experiment was the Agfa–Gevaert holographic plate with an intrinsic resolution of $1.5\,\mu m$.

Figure 3.15 shows the Fourier image of a copper grid, recorded in the system illustrated in figure 3.13. The image magnification is about 1.5, the image cell

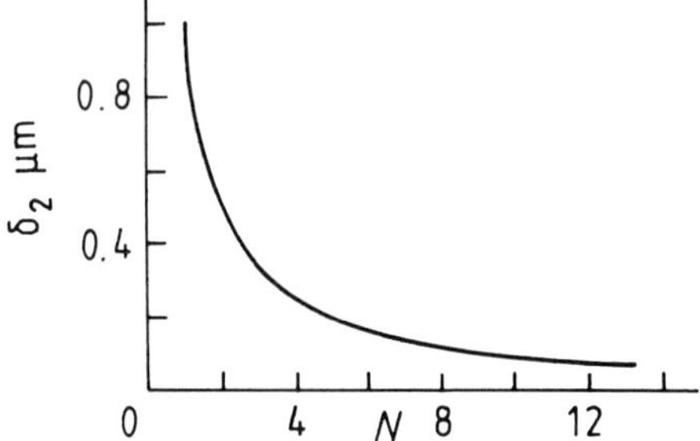

Figure 3.12 Limiting resolution of Fourier images as a function of the number N of mask periods and $d_M = 1\,\mu\text{m}$.

size is approximately $20\,\mu\text{m}$ and the resolution is of the order of $5\,\mu\text{m}$. The second-order Fourier image is more complicated (figure 3.15(*b*)) because it is an interference pattern due the superposition of cell images.

Much more powerful sources of radiation and higher detector resolution are required to produce two-dimensional Fourier images with the resolution of better than $1\,\mu\text{m}$ that is required, for example, for lithographic purposes [113]. Such experiments were first published in [114–116].

Figure 3.16 illustrates the experiment performed on the Photon Factory at Tsukuba, Japan [115]. The source was an undulator mounted in the beam line of the storage ring of the photon factory [117]. The experiments were performed using an electron beam current of 80–$100\,\text{mA}$ and two stops with aperture diameters of $1\,\text{mm}$ were inserted into the X-ray beam. The first was at a distance of $10\,\text{m}$ and the second at $5\,\text{m}$. The hard component of the radiation was cut off by a silicon mirror, mounted so as to produce total external reflection. The sample and the detector were held in a special device in which the separation between the object and the silicon plate coated with the X-ray resist could be accurately monitored. The working wavelength ($2.4\,\text{nm}$, near the K-absorption edge of oxygen) was chosen because polymer substrates had to be used for the X-ray templates. The best resolution expected in this experiment was estimated from (3.19) and (3.23).

A specially prepared X-ray template on a polyimide membrane with a high-contrast gold pattern was used in this experiment (figure 3.17(*a*)). The period of the structure was $5\,\mu\text{m}$ and the line width $0.3\,\mu\text{m}$. Figure 3.17(*b*) shows the Fourier image on a X-ray resist, written with this X-ray template using a wavelength of $2.4\,\text{nm}$ in undulator radiation. The object–detector separation was $10\,\text{mm}$ and the expected resolution was $0.25\,\mu\text{m}$. The results obtained are in good agreement with these estimates. Figure 3.18 shows a number of Fourier images of the copper grid illustrated in figure 3.14. The 16th-order Fourier image shows details with dimensions of $0.2\,\mu\text{m}$ (figure 3.18(*c*)).

3.2.2 Fourier imaging in incoherent radiation

The basic requirement in the production of high-resolution Fourier images is

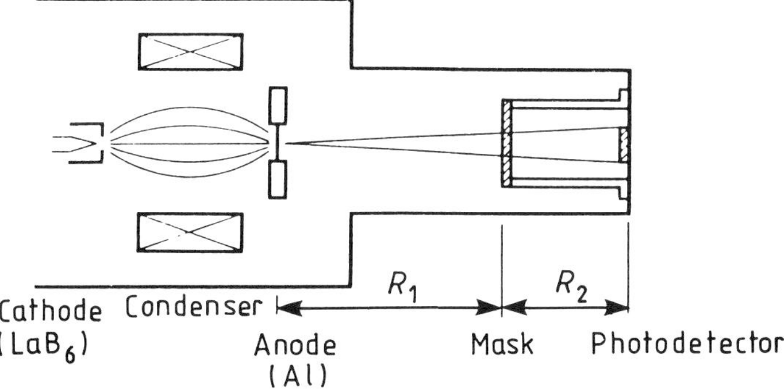

Figure 3.13 Schematic diagram of the aparatus used to observe Fourier images with X-ray microfocus source (wavelength 0.83 nm).

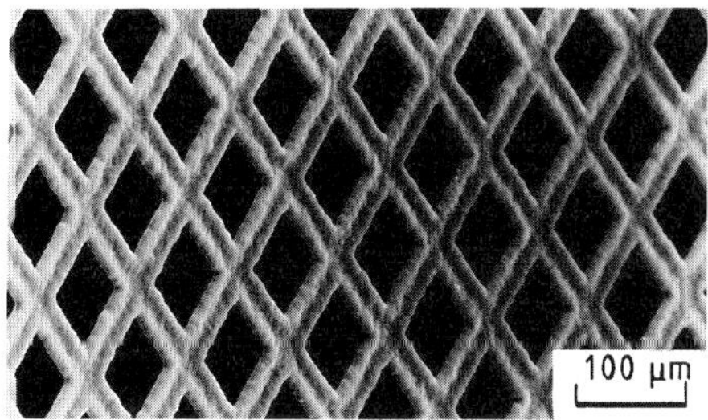

Figure 3.14 Micrograph of a copper grid with a period of 12.7 μm. The grid was used to produce Fourier images shown below.

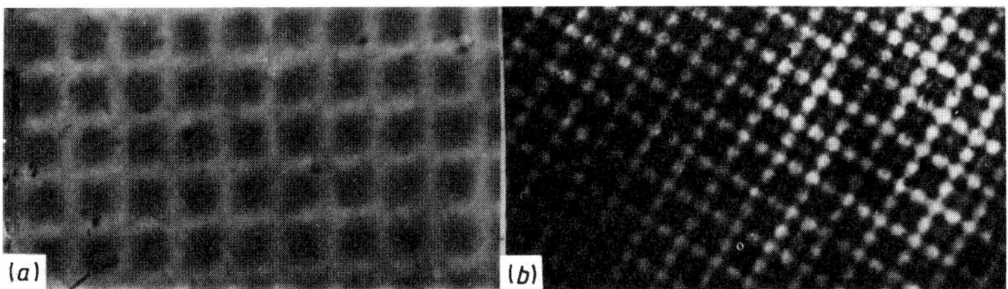

Figure 3.15 Fourier image of a copper grid recorded in the $m = 1$ (a) and $m = 2$ (b) Fourier planes (radiation wavelength 0.83 nm).

that the illuminating radiation must be very coherent both in space and in time [111]. An increase in source size, or a departure from monochromaticity, produces a defocusing of the image. However, the analytical properties of the coherence function and, in particular, the van Cittert–Zernike theorem for a plane source with a periodic brightness distribution, can be exploited to produce high-

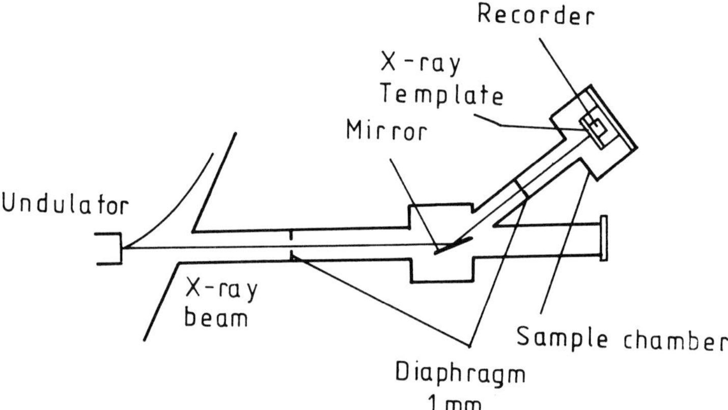

Figure 3.16   Diagram of aparatus used to observe Fourier images in undulator radiation [115].

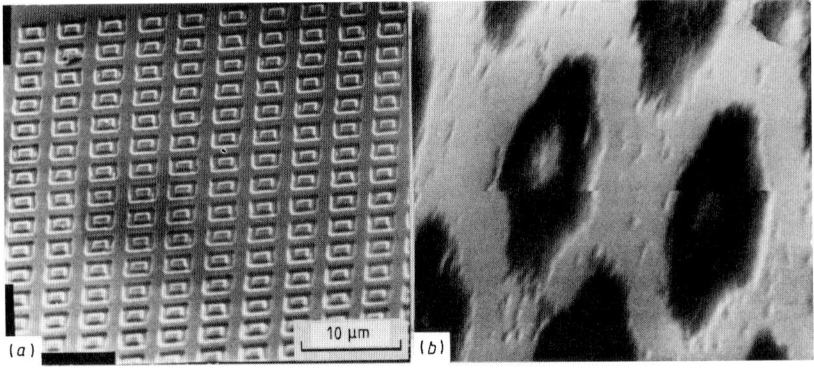

Figure 3.17   Micrograph of a polymer X-ray template with a gold mask (a) and its Fourier image at 2.4 nm for a mask-detector separation of 10 mm (b) [115].

resolution Fourier images in white light with a source of radiation of arbitrary size [118–120]. In the case of Fraunhoffer diffraction, the analogue of this process is the Lau effect [107–109] which can be used to obtain the spectrum of Fourier images.

An optical system based on the Talbot effect and using modulated spatial coherence of radiation is discussed in [112, 120] in the context of coherent and incoherent image transfer. It is shown that a high-resolution image can be produced even in white light.

Figure 3.19 illustrates the optical arrangement employing modulated spatial coherence. Totally incoherent radiation is used to illuminate an object after passing through an array of secondary sources (periodic mask with apertures).

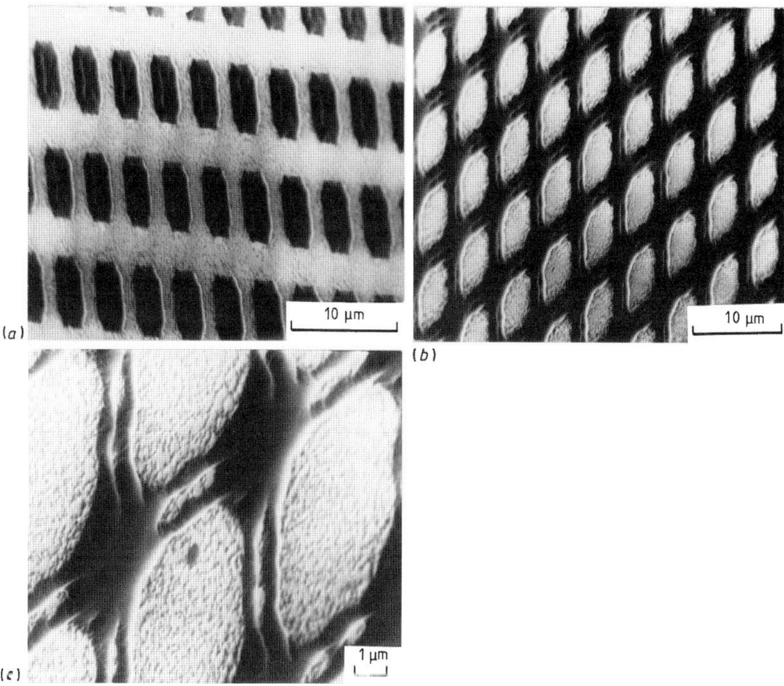

Figure 3.18 Micrographs of Fourier images obtained for the following distances from the mask: 64.5 (*a*), 32.3 (*b*) and 4 mm (*c*).

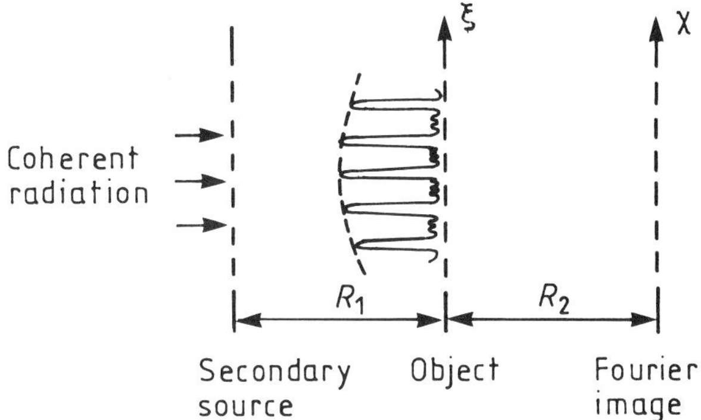

Figure 3.19 Method of producing a Fourier image in radiation with modulated spatial coherence.

According to the van Cittert–Zernike theorem, the spatial coherence function of the radiation is the Fourier transform of the intensity distribution in the plane of the secondary sources [120]. If we assume that v is the position coordinate in the plane of the source mask (x in the object plane and u in the image plane), the spatial coherence function in the image plane can be described as follows. The intensity distribution in the plane of the entrance transparency, i.e., the source matrix, is

$$f(v) = \sum_{h=-(H-1)/2}^{h=(H-1)/2} \delta(v - hd_M) * g_s(v) \tag{3.25}$$

where d_s is the source period and $g_s(v)$ is the transmission function of a line. According to the van Cittert–Zernike theorem [123], the spatial coherence function in the object plane is

$$\Gamma(\Delta x) = \exp\left[\frac{i\pi}{R_1\lambda}\left(x_1^2 - x_2^2\right)\right] G_s\left(\frac{\Delta x}{R_1\lambda}\right)$$
$$\times \sum_h \delta\left(\frac{\Delta x}{R_1\lambda} - \frac{h}{d_s}\right) * Hd_s \operatorname{sinc}\left(\pi Hd_s \frac{\Delta x}{R_1\lambda}\right) \tag{3.26}$$

where $\Delta x = x_1 - x_2$, $G_s(\Delta x)$ is the Fourier transform of $g_s(v)$, H is the number of apertures in the source mask and R_1 is the distance between the source mask and the object.

According to [48] the intensity in the image plane can be written in the form

$$I(u) = \frac{1}{R_2^2\lambda^2} \int\limits_{S_1} \int\limits_{S_2} \Gamma(\Delta x) \exp\left\{ i\frac{\pi\left[(x_1 - u)^2 - (x_2 - u)^2\right]}{R_2\lambda} \right\} dx_1\, dx_2 \tag{3.27}$$

where R_2 is the object–image separation. The integral is evaluated over S_1 and S_2 (the surfaces of the object mask). If the object mask is a periodic amplitude transparency, we can discard the constant factors in (3.27), so that

$$I(u) \approx \int\limits_{S_1} \int\limits_{S_2} g(x_1)g(x_2)\Gamma(\Delta x)E(x_1 - u)E(x_2 - u)\, dx_1\, dx_2. \tag{3.28}$$

where $g(x)$ is the amplitude transmission function of the object and

$$E(x - u) = \frac{1}{R_2\lambda} \exp\left[\frac{i\pi}{R_2\lambda}(x - u)^2\right]$$

Let us now consider two limiting cases of the solution of (3.28).

First, for completely coherent radiation (plane wave), $\Gamma(\Delta x) = 1$ and the intensity distribution in the u plane is

$$I(u) = \left\{ \int g(x_1)\, E(x_1 - u)\, dx_1 \right\}\left\{ \int g(x_2)\, E(x_2 - u)\, dx_2 \right\}. \tag{3.29}$$

This is analogous to (3.15) and describes image transfer in coherent radiation. Since the function $g(x)$ is specified on a finite aperture Hd_M, the expression given by (3.29) takes the form

$$I(u) \approx \left| g(u) * \mathrm{sinc}\left(\pi H d_M \frac{u}{R_2 \lambda} \right) \right|^2 . \tag{3.30}$$

Second, the source of radiation is completely spatially incoherent and takes the form of a periodic sequence of narrow slits. This means that (3.26) can now be written in the form

$$\Gamma(\Delta x) \approx \exp\left\{ \frac{i\pi}{R_1 \lambda} \left(x_1^2 - x_2^2 \right) \right\} \sum_{-(H-1)/2}^{(H-1)/2} \delta\left(\frac{\Delta x}{R_1 \lambda} - \frac{h}{d_s} \right) \tag{3.31}$$

so that, evaluating the integral with respect to x_2 in (3.28) and assuming that the synchronisation condition

$$R_1 \lambda = d_M d_s \tag{3.32}$$

is satisfied, we have

$$I(u) = \sum_{-(H-1)/2}^{(H-1)/2} \int_S g(x) g(x + h d_M) E(x - u) E^*(x + h d_M - u) \, dx. \tag{3.33}$$

Since $g(x) = g(x + H d_M)$, and the aperture of the object mask is finite, we have

$$I(u) \approx H g(u)^2 * \mathrm{sinc}^2\left(\pi H d_M \frac{u}{R_2 \lambda} \right). \tag{3.34}$$

The image intensities H due to the sequence of slits are summed in the image plane. We note that high-contrast images can be produced in this case by using a spatially incoherent source with a white spectrum. The phase factor $\exp[i\varphi(x)]$ produces image magnification in accordance with the lens formula (3.24). The image is transferred by analogy with an objective of diameter $H d_N$ and focal length F_m in a spatially incoherent radiation field. A more detailed discussion of the transfer function of a system in the case of coherent and incoherent image transfer is given in [120].

Incoherent image transfer has the remarkable advantage that it is not accompanied by interference noise from different diffraction orders. Figure 3.20 shows two images recorded in the case of spatially coherent (small source) and incoherent (array of secondary sources) image transfer. A wide-spectrum source of white radiation was used in both cases. The spatial resolution of a system producing incoherent image transfer based on the Talbot effect is measured by the square of the ratio of the slit width in the array of secondary sources to

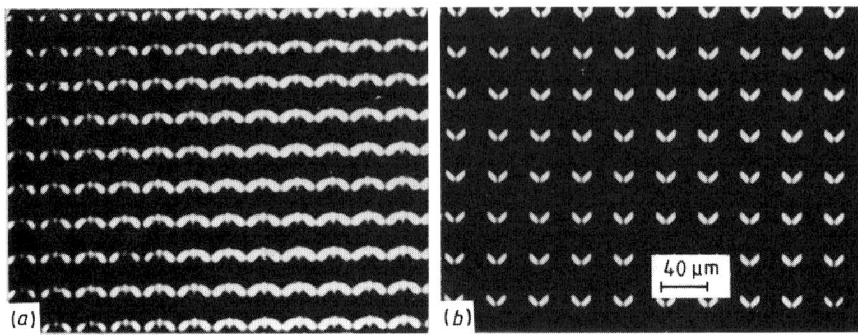

Figure 3.20 Image of a two-dimensional object produced in white light, using incoherent transfer of the Fourier image [112]: (*a*) spatially coherent source (point source), (*b*) incoherent source.

the effective period of the structure. Actually, in the case of image defocusing, the minimum size of the focal spot is determined by the width of the effective spectrum of incident radiation $\Delta\lambda_{\text{eff}}$ and is given by

$$\delta_x \approx \sqrt{\Delta\lambda_{\text{eff}}\,(R_1 + R_2)} \tag{3.35}$$

where $R_1 + R_2 = [d_M^2 + d_M d_s]/\lambda$.

Let us suppose that

$$d_{\text{eff}}^2 = d_M^2 + d_s d_M \tag{3.36}$$

is the effective period of the entire Talbot optical system. The effective relative width of the spectrum producing the image can be obtained from (3.35) and is given by

$$\Delta\lambda_{\text{eff}}/\lambda = (\delta_x/d_{\text{eff}})^2 .$$

In the incoherent Talbot system, the image is transferred by point gratings with line width b_s, so that the effective spectrum width and, therefore, the resolution, are given by

$$\Delta\lambda_{\text{eff}}/\lambda = b_s^2/d_{\text{eff}}^2 .$$

We therefore conclude that, as $b_s \to 0$, each maximum (each image point), is formed from quasimonochromatic radiation with high resolution. The quantity b_s is limited by the reduction in image contrast, i.e., the superposition of defocused images produced in other wavelengths. There is therefore an optimum radiation line width that produces given resolution for an acceptable image contrast. Numerical simulations show that, for example, when the size of the structure element is $20\,\mu$m and the period is $400\,\mu$m, the optimum spectrum width of a source at a wavelength of $0.5\,\mu$m should be $\Delta\lambda/\lambda = 0.25$ for an image contrast of the order of 5%.

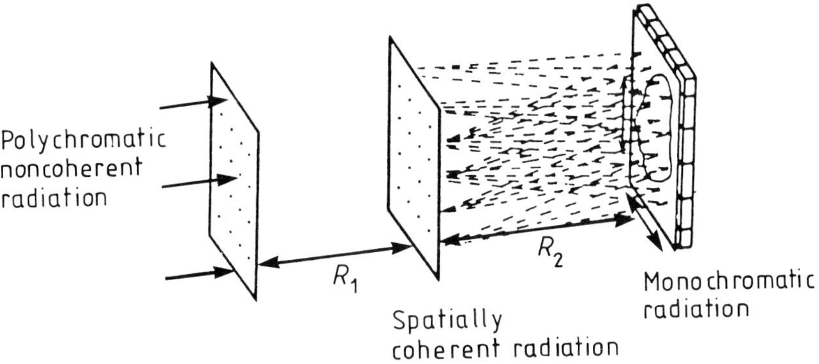

Figure 3.21   Principle of the scanning microscope with a Talbot objective [112].

3.2.3   Diffraction objective based on the Talbot effect

A system employing a mask with point apertures for the transfer of images of nonperiodic objects in monochromatic radiation was proposed in [121]. In this arrangement, each object point in the source plane produces in the image plane a Fourier image of the mask consisting of point apertures, and these images add incoherently to the Fourier images due to other object points.

The pinhole mask is a diffraction objective with a set of focal lengths given by $n, m = 1, 2, ...$, where n is the number of the Fourier plane without period multiplication and m is the number of period multiplications in the given Fourier plane. The expression relating the separation between the objective and the object to the separation between the objective and the image is analogous to the lens formula (see (3.24)). If the object is a single non-periodic structure, R_1 and R_2 can vary continuously and the magnification of the objective $M = (R_1 + R_2)/R_1$ varies continuously between zero and infinity. On the other hand, if the object is a periodic structure, then in addition to (3.6) we must satisfy the further condition that the Fourier images of the pinhole mask due to object points separated by one period must coincide. Consequently, the distances R_1 and R_2 and the magnification M must change discretely.

In addition to the projection optical system, the Talbot objective can also be used in a scanning arrangement (figure 3.21). This scheme is convenient in X-ray microscopy because the multi-focus Talbot objective can scan simultaneously several portions of the object. This results in a substantial reduction in the image formation time.

At first inspection, the objective resembles the *pinhole camera*. Indeed, the magnification M of this objective and the resolution $\delta_x = b'_s(R_1 + R_2)/R_1$ is the same as for the pinhole camera. However, the principles underlying the operation of the two devices are quite different. The pinhole camera is based entirely on the shadowing principle of image formation. It works so long as the size of the diffraction spot $\delta_{\text{diff}} = \lambda R_2 / b_s$ does not exceed the magnified shadow image of the aperture $b'_s = b_s (R_1 + R_2)/R_1$. There is no point in reducing b_s any further in the pinhole camera, i.e., there is no point in increasing the resolution beyond this point, since the size of the diffraction spot would then become greater than b'_s. The condition for the pinhole camera is therefore

$$b_s^2 > \lambda R_2 R_1 / (R_1 + R_2). \tag{3.37}$$

This also determines the maximum resolution of the pinhole camera. For example, for $\lambda = 5\,\text{nm}$, $R_1 = 1\,\text{cm}$ and $R_2 = 100\,\text{cm}$, the minimum size of the aperture is $6\,\mu\text{m}$.

The Talbot objective relies on the phenomenon of diffraction and begins to work when the aperture size reaches the 'difraction value'. The working condition for the Talbot objective is therefore

$$b_s^2 < \lambda R_2 R_1 / (R_1 + R_2) \tag{3.38}$$

so that the resolution is practically unlimited as b_s is reduced. The diffraction limit of resolution $\delta_{\text{diff}} = \lambda R_2 / H d_s$ is determined by the total aperture of the Talbot objective.

3.2.4 Talbot-effect spectrometry

As noted above, the focal length of the Talbot objective is a function of wavelength. This means that, as we reduce the distance from the objective, the Fourier images are formed by different wavelengths. The effect is therefore spectrally selective, and we can speak of the detection of the emission spectrum under the conditions of Fresnel diffraction. The spectrum is recorded by varying the object–image separation. It follows from (3.8) that the longitudinal dispersion is given by

$$\Delta R / R_2 = \Delta \lambda / \lambda. \tag{3.39}$$

Let us now estimate the resolving power of this spectral device. Suppose the object is the usual periodic grating with slit width b_s. The image is given by the convolution of the intensity distribution in the object plane and the influence function of the optical system, $H(\nu)$. The latter describes the diffraction spreading of the image. The defocusing in the region in which Fresnel diffraction takes place is known to be

$$\delta_z \approx \sqrt{\Delta R \, \lambda} \tag{3.40}$$

and is a measure of the depth of focus ΔR. By analogy with the Rayleigh criterion, the resolving power can be estimated by equating the depth of focus to the separation between two closely spaced wavelengths: $\Delta \lambda / \lambda = \Delta R / R$. The depth of focus and, consequently, the resolving power are determined by two factors, namely, the width of the diffraction maximum of the influence function and the slit size. Substituting $\delta_z = \lambda R/(H d_M)$ in (3.40) and recalling the relation $\lambda = d_M^2/R$, we find that the resolving power due to the diffraction defocusing of the image is given by

$$\Delta\lambda/\lambda = 1/H^2 \qquad (3.41)$$

where H is the number of slits in the grating.

We recall that, when the spectrum is recorded under the conditions corresponding to Fresnel diffraction, the resolving power is $\Delta\lambda/\lambda = H^{-1}$ and does not depend on the slit width. This result was obtained in [122], but the authors of this paper ignored the more important second factor. In practice, the slit size cannot be reduced to the size of the diffraction maximum because of the sharp reduction in transmission. Substituting $\delta_{\text{diff}} = b_s$ in (3.40), we find that the resolving power due to the finite slit width in the diffraction grating is

$$\Delta\lambda/\lambda \approx (b_s/d_M)^2 . \qquad (3.42)$$

The resolving power is thus proportional to the square of the 'duty ratio' of the grating. The yellow doublet in the spectrum of mercury vapour ($\lambda_1 = 579\,\text{nm}$, $\lambda_2 = 577\,\text{nm}$) was resolved in the model experiment reported in [122] by using a grating in which the number of slits was $H = 20$. We note that the grating duty ratio $d_M/b_s > 20$ was not sufficient to resolve this doublet. A grating with the number of slits in excess of 300 was necessary to resolve the doublet (the mark-space ratio is unimportant here).

It is thus clear that optical systems based on the Talbot effect and using modulated spatial coherence can be exploited to produce high-resolution images in non-monochromatic light. Spatially incoherent image transfer gives rise to interference patterns in the component wavelengths, which distort the images of the object. They disappear in the case of incoherent image transfer. The Talbot diffraction objective discussed in [112] has the properties of the more conventional image forming systems, but can also be used at shorter wavelengths for which there are no analogues of optical elements. The dispersive properties of the Talbot system can be used in high-resolution spectroscopic studies with less stringent conditions imposed on the aperture of the dispersing grating.

4

Synthesized three-dimensional Bragg–Fresnel X-ray optics

This chapter presents the principles and properties of three-dimensional focusing and dispersive elements based on multilayer mirrors and known as Bragg–Fresnel elements. We begin by analysing the diffraction principles underlying the operation of the multilayer X-ray mirrors, and the rules governing the choice of materials and fabrication technology. The spatial structure and properties of Bragg–Fresnel lenses (BFLs) are analysed in ellipsoidal coordinates. The kinematic approximation is used to estimate the efficiency and spectral selectivity of the BFL. Next, a kinematic model is presented together with experimental results obtained for X-ray acoustic modulators–deflectors based on multilayer structures modulated by a surface acoustic wave. The chapter concludes with a comparison and detailed analysis of the properties of planar and Bragg–Fresnel X-ray optical elements.

4.1 MULTILAYER X-RAY MIRRORS

Some years after the discovery of X-rays, it was suggested that multilayer structures consisting of elements of different atomic number could be used to reflect radiation in the X-ray range. Bragg diffraction by natural crystals is an example of this type of reflection. Soon after it was demonstrated that X-rays were in fact waves, and a diffraction theory was developed for them, it was suggested that such structures could be fabricated artificially, using layers of thickness ranging from a few units to a few dozen ångströms. The first attempts to create such mirrors by evaporation and electro-deposition were reported in [123]. However, X-ray diffraction by the superlattice of such artificial layers was not observed at the time.

The first successful use of multilayer mirrors was reported in [124]. A periodic structure was produced by depositing copper and gold layers with a period of 10 nm, and this produced the diffraction of Mo Kα X-rays ($\lambda = 0.07$ nm) by the superlattice. However, the stability of the mirror remained a

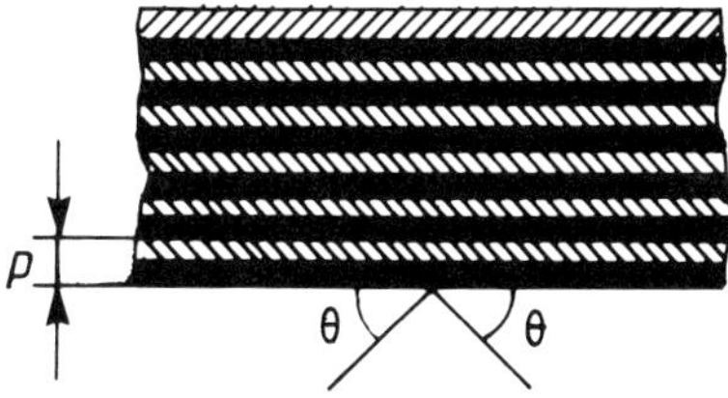

Figure 4.1   A multilayer X-ray mirror.

problem. It was found that the diffraction efficiency fell to zero over the period of a month because of the interdiffusion of the metals and the reduction in the difference between the refractive indicies of the two layers. For a long time, this presented a constant difficulty for multilayer technology. It took 25 years before it was found [125,126] that the iron–magnesium multilayer structure with a period of 3–5 nm retained its diffraction properties for about a year. During this year the OKα X-ray diffraction peak ($\lambda = 2.36$ nm) fell in intensity in the same way as for lead stearate films produced by the same technology that were used for Langmuir–Blodgett of seven films.

Multilayer coatings for the vacuum ultraviolet were subsequently produced [127–129] by the evaporation methods. These coatings do not constitute a periodic structure in the usual sense of this phrase. Although the sum of the thicknesses of the two layers remains constant, their relative thickness is varied to produce maximum reflection for a large number of layers. Measurements on synchrotron radiation have shown that copper–carbon and gold–carbon layers have reflection coefficients at the wavelength of 19.0 nm that are higher by a factor of seven as compared with a thick layer of gold at normal incidence.

For angles of incidence much greater than the angle of total external reflection, the X-ray intensity reflection coefficient does not exceed 10^{-4}–10^{-6}. However, the corresponding amplitude coefficient is 10^{-2}–10^{3}. This means that a reflection coefficient of the order of unity can be achieved by using suitably phased reflections from 100–1000 layers. Synthesized multilayer mirrors are based precisely on this principle and the theory of diffraction by multilayer coatings is developed in [129–135].

Consider a multilayer mirror consisting of two materials A and B (figure 4.1) of thickness d_A and d_B, respectively. The period of the structure at right angles to the two layers is therefore $d = d_A + d_B$. The results that are obtained for this case can readily be generalized to a large number of different media. X-ray radiation of wavelength λ, incident at an angle θ is scattered by each layer of atoms and reflections are found to be in phase when the Bragg condition

$$m\lambda = 2d \sin \theta_m$$

is satisfied where m is the diffraction order.

Multilayer mirrors can be classified into two basic types depending on the optical properties of media A and B and, consequently, their thicknesses:

(1) A combination of a very thin layer of a heavy metal and a carbon interlayer (for example, tungsten and carbon) $d_A \ll d_B$.

(2) A combination of two relatively light materials with similar properties, (for example, nickel and carbon): $d_A \approx d_B$; improved spectral properties can be achieved in this case.

The Bragg-diffracted intensity can be calculated for a multilayer mirror by solving the Maxwell equations for a medium with a periodically varying refractive index. The problem is completely analogous to the determination of the Bragg-scattered intensity for a crystal structure and can be solved approximately in two cases. In the kinematic theory, scattered radiation is received independently from each volume element. This theory neglects the reduction in the amplitude of the direct wave as it propagates in the structure (extinction). All interactions are taken into account in the dynamic theory. This theory is usually employed to describe diffraction in perfect crystal of considerable thickness whereas the kinematic theory is more suitable for 'thin' or 'mosaic' crystals.

Multilayer mirrors are perfect from the crystallographic point of view. Even variations in the layer thickness introduced by the fabrication process are not equivalent to the mosaic structure of natural crystals. Consequently, the dynamic theory is more suitable for the description of diffraction by multilayer mirrors with high reflection coefficients [132].

Absorption in a layer can be taken into account by introducing a complex refractive index $\hat{n}$. A good approximation is provided by the theory developed in [136] for three-dimensional reflecting holograms (coupled-wave theory) or by numerical calculations of diffracted amplitudes. A computational model that takes into account absorption is presented [134] together with examples of the angular distributions produced by carbon–tungsten structures with different numbers of layers.

A very important fact that determines the diffraction efficiency of multilayer mirrors is the quality of the surface, and consequently, the uniformity of the layers [137, 138].

The durability and efficiency of multilayer structures depend on the choice of the media for them. The problem consists, in effect, of two parts, namely, the identification of the properties of thin films, e.g., their interdiffusion rates and the examination of the X-ray optical properties of the media [139–143].

Stable multilayer structures include carbon–tungsten, carbon–vanadium, carbon–titanium [130], carbon–nickel, carbon–iron. W–Re/C, W–Re/Si and Si/Al are used in the vacuum ultraviolet. Au–Pd/C and Re–W/B are investigated in [129]. Figure 4.2 shows the limiting number of periods for multilayer mirrors with carbon as one of the layers. The dashed line shows the actual values for Ni/C.

Evaporation by electron-beam bombardment is the traditional method used

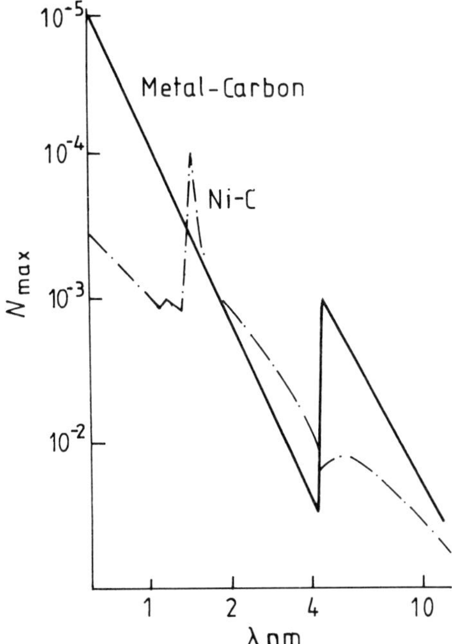

Figure 4.2 Maximum number of layers in a multilayer mirrors containing carbon films [142]. The dashed line shows the actual values for the C–Ni combination.

to produce multilayer coatings [129,144]. The essential part of the system described in [129] is the reflectometer working in the soft X-ray region. It is used to monitor the thickness of the deposited layer by examining the reflection produced by it. The interference maxima and minima recorded during the deposition process provide an accurate measure of the thickness and are used to eliminate drifts that are possible in the case of calibration against quartz probes. The main advantage of this arrangement is that errors in the growing film do not accumulate and the maximum total error is a small fraction of the period of the structure. The $K\alpha$ lines of carbon, nitrogen and boron (4.47, 3.16 and 6.76 nm, respectively) are usually used to monitor the mirror thickness. Polished silicon or glass are used as substrates.

A magnetron method with a fixed source and a moving substrate is used in [130]. Here all the sources at an equal distance from a centre and the stage upon which the substrates is mounted is rotated under the sources. This method minimizes the influence of substrate inhomogeneities and produces layers of thickness up to several ångströms.

The laser method of depositing multilayer mirrors has recently undergone considerable development [139,145,146]. It produces high-grade films of very reproducible thickness. The laser method can readily be automated and is

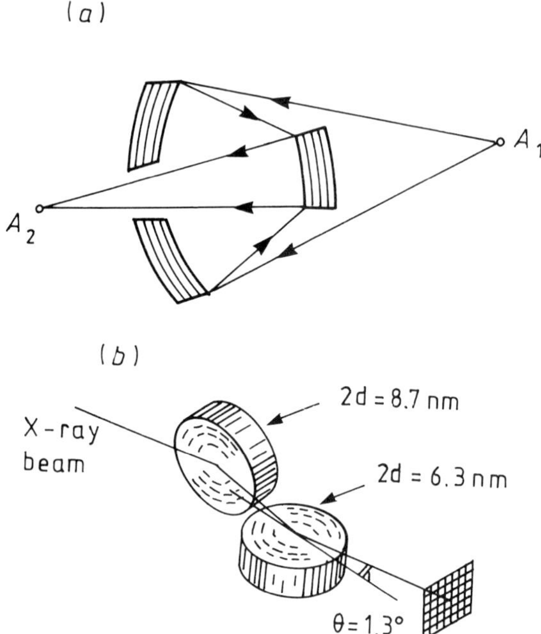

Figure 4.3 (*a*) The Schwarzschild objective employing multilayer mirrors [150] and (*b*) the Kirkpatrick–Baez microscope incorporating multilayer spherical mirrors [151].

probably one of the most promising techniques for applications in industry [138,145,146].

Multilayer mirrors have found extensive applications as spectral and focusing elements in X-ray optics [147–149]. The deposition of multilayer coatings on the reflecting surfaces of a particular shape can be used to produce precision focusing and imaging systems as well as normal-incidence components for the X-ray range. By replacing grazing-incidence mirrors in the Wolter objective with multilayer mirrors, it is possible to investigate the soft component of radiation in astrophysical and plasma-physics experiments [147]. Multilayer mirrors have been used to develop an X-ray version of the Schwarzschild objective at a wavelength of 30.4 nm [150]. This is illustrated in figure 4.3(*a*).

An original electron-beam fabrication techology employing a rotating mask was used to produce the mirrors for this objective. This procedure was used successfully to fabricate mirrors on a spherical surface with a variable period, which ensured a large effective aperture of the objective. A single spherical mirror, used as the focusing element, does not produce satisfactory resolution because of the attendant large off-axis aberations (essentially distortion). A resolution of the order of 1 μm at a wavelength of 0.15 nm can be achieved by using two spherical mirrors in the Kirkpatrick–Baez scheme (figure 4.3(*b*)). The theory of ellipsodial focusing multilayer structures is presented in detail in [152]. However, the problem of multilayer mirrors in high-resolution X-ray microscopy cannot be solved by using curved profiles because it is practically impossible to achieve a resolution of this method of $1 \, \mu$m. In the next section, we shall discuss the principles of another type of optics based on multilayer mirrors, namely, Bragg–Fresnel optics.

4.2 SPATIAL STRUCTURE AND THE PROPERTIES OF THE BRAGG–FRESNEL LENS

Simultaneous monochromatization and focusing of incident radiation is possible by creating a diffracting element on a X-ray mirror. This type of element is often called a Bragg–Fresnel lens [153] or in some papers an 'enhanced dispersion grating' [154,155]. In the general case, the surface profile of multilayer mirrors is quite complicated.

4.2.1 Diffraction model of a Bragg–Fresnel lens

The spatial structure of a Bragg–Fresnel optical element can be described quite simply in terms of the interference pattern produced by two spherical wavefronts (figure 2.1) which provides the basis for the description of diffraction optics. The ideal system is probably a structure that completely reproduces the interference pattern profile as in the case of three-dimensional reflecting holograms. One way of achieving this is to use flat or curved substrates with the equal-phase Fresnel zones enscribed upon them. This type of element is shown schematically in figure 4.4 in which a multilayer structure cuts the Fresnel zones and forms the three-dimensional structure of the Bragg–Fresnel element. Such elements are produced in practice by constructing a spatial model and developing algorithms for computing the necessary sections.

The spatial structure is conveniently described in terms of ellipsoidal or parabolic coordinates. The set of spatial Fresnel zones is a family of confocal ellipsoids or paraboloids of revolution for which the sum of focal lengths differs by $\lambda/2$ (see figure 2.1). Let us suppose that the family of confocal ellipsoids is described by

$$\frac{x^2}{A_1^2} + \frac{y^2}{A_2^2} + \frac{z^2}{A_3^2} = 1. \tag{4.1}$$

We now transform to ellipsoidal coordinates [54] defined by

$$x = a\xi\eta \qquad y^2 = a^2\left(\xi^2 - 1\right)\left(1 - \eta^2\right)\sin^2\varphi$$
$$z^2 = a^2\left(\xi^2 - 1\right)\left(1 - \eta^2\right)\cos^2\varphi \qquad 0 < \eta < 1 \qquad \xi \geqslant 1 \tag{4.2}$$

where ξ, η, φ are the ellipsoidal coordinates and a is the focal parameter of the family of ellipsoids. We also assume that the sections of the ellipsoids are

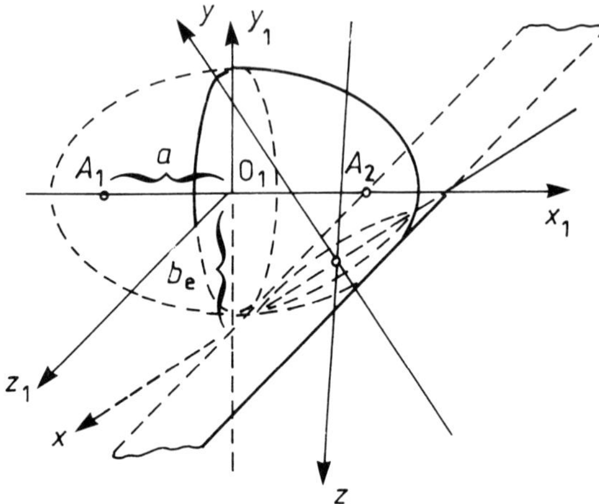

Figure 4.4 Sections of surfaces of equal phase by a multilayer structure: A_1 source of radiation, A_2 focal point, a focal parameter, b_l segment cut by layer l on the Y axis, X_1, Y_1, Z_1 original coordinate frame, X, Y, Z coordinate frame in the plane of the section.

parallel to the Z axis with a slope k to the X axis:

$$z = kx - b_l \qquad b_l = b_0 + l d \sqrt{k^2 + 1} \tag{4.3}$$

where l is the layer number. After rotation and shift of coordinates, we obtain

$$x = \left(1 + k^2\right)^{-1/2} \left[a\xi\eta\left(1 + k^2\right) - b_l k\right] - \frac{ka^2}{b_l \sqrt{k^2 + 1}} \tag{4.4}$$

$$y = -b_l \left(1 + k^2\right)^{-1/2} \tag{4.5}$$

$$z = \pm \left[a^2 \left(\xi^2 - 1\right)\left(1 - \eta^2\right) - (ka\xi\eta - b_l)^2\right]^{1/2}. \tag{4.6}$$

If ξ_0 and η_0 are the coordinates at which the 'reference' ellipsoids touches the surface of the multilayer structure, we find that the condition for maximum intensity at the focus after reflection from any of the surfaces is

$$\xi_0 + 2n\lambda/(4a) \leqslant \xi_n \leqslant \xi_0 + (2n + 1)\lambda/(4a). \tag{4.7}$$

Ellipsoids with the parameters ψ_0 satisfying this condition contribute to the focus with the same sign of the phase. It is therefore a relatively simple matter to use (4.4)–(4.7) to find the coordinates of the intersection of the 'Fresnel zones' with the surface of the multilayer structure.

In the general case, the zone profile is an ellipse with a shifted centre, i.e., a closed second-degree curve. This shape of the section imposes special requirements on the supplication technology for such structures. Figure 4.5(*a*)

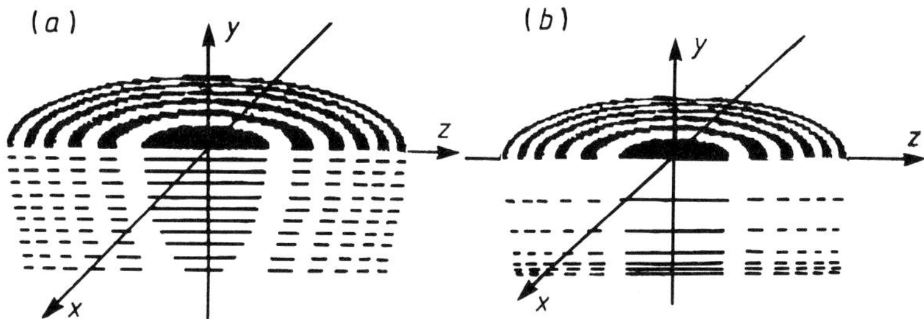

Figure 4.5 The three-dimensional structure of the Bragg–Fresnel lens with fixed (a) and variable (b) period [165].

shows the profile of three-dimensional Fresnel zones within a multilayer structure. It shows a model of a transverse section of a multilayer mirror with equidistant layers of reflecting material. It is clear from figure 4.5(a) that the zone depth profile is also complicated in shape. We note that this zone shape is the exact solution and determines the interference focusing of radiation in the Fresnel diffraction region.

The closest analogue to the Bragg–Fresnel focusing system is the three-dimensional hologram whose properties are those of three-dimensional diffraction [156, 157]:

(1) The resolving power and the shape of the diffraction maximum are determined by the three dimensional Laue function.

(2) Each surface reflects the wave incidence upon it towards the focal point A_2 independently of the wavelength λ, so that the ideal Bragg–Fresnel length (BFL) does not suffer from chromatic aberration.

(3) The field of view of the lens is small and is determined by its aperture. The separation between the source of radiation and the lense along the OA_1 access can be varied between relatively wide limits. As in ordinary optics, the standard lens formula is valid.

(4) A change in the separation of the reflecting planes by the amount Δd in neighbouring portions of the lens produces a change in the phase of the waves scattered by these portions by an amount proportional $\Delta d/d$ where d is the separation between the planes. In principle, this should allow rapid manipulation of reflected beams, e.g., fast switching.

In most experimental situations, reflections by the Bragg–Fresnel lens cannot apparently be described in the first Born approximation and must be calculated with allowance for multiple-reflection effects. However, in the soft X-ray range, the number of layers in the the multilayered structure does not exceed 100 so that diffraction processes, can be qualitatively described in terms of the kinematic approximation, which will be done, for example, in the case of modulation of X-ray beams.

The three-dimensional distribution of Fresnel zones is significantly different when the layers in the multilayer structure are not arranged equidistantly. When the period of the layers is allowed to vary in accordance with the expression

$$d_l = a \left[\left(\xi_0 + \frac{l\lambda}{2a} \right)^2 (k^2 + 1) - 1 \right]^{1/2}$$
$$- a \left[\left(\xi_0 + \frac{(l-1)\lambda}{2a} \right)^2 (k^2 + 1) - 1 \right]^{1/2} \tag{4.8}$$

the distribution of the zone structure within the layer is shown by figure 4.5(b). The zones tend to straighten out in the vertical direction which is exceedingly important from the point of view of the practical implementation of the Bragg–Fresnel optical principle. The shape of the zones lying near the edge of the Bragg–Fresnel element is particularly important.

The focusing properties of the Bragg–Fresnel optical lens are determined by the dimensions of its diffracting elements just as in the case of ordinary Fresnel zone plate.

Properties of Bragg–Fresnel optical elements. The Bragg–Fresnel lens is a three-dimensional diffracting structure with well defined selective properties in relation to both the angular and energy distributions in the incident wave. In the case of multilayer X-ray mirrors, the precise magnitude of the reflection coefficient of the multilayer structure can only be obtained by using the dynamic theory. However, a qualitative description that reproduces the overall picture and is sufficient for approximate estimates can be performed in the kinematic approximation, i.e., by ignoring the influence of diffraction on the incident wave [157]. For the sake of simplicity, let us consider the one-dimensional case.

Let x be the running coordinate on the zone plate and $y = di$ be the coordinate measured along the normal to the multilayer structure. The wave amplitude diffracted by the multilayer BFL can be written in the form

$$E(x, y) = V \sum_{l=1}^{L} \int_{S_l} \exp\left[i\frac{2\pi}{\lambda} (R + r) \right] dx \tag{4.9}$$

where V is the Fresnel reflection coefficient of a layer,

$$V = \frac{\cos\theta_0 - (1 - \delta)\cos\theta_1}{\cos\theta_1 + (1 - \delta)\cos\theta_0}. \tag{4.10}$$

The radii R_1 and R_2 can be obtained by using the approximate expansion

$$R = R_1 - xx_1/R_1 - yy_1/R_1 \tag{4.11}$$

where $R_1^2 = x_1^2 + y_1^2$ is the square of the distance to the source of radiation (we are assuming that the source of radiation is at a large distance from the optical element, so that the quadratic terms in (4.11) can be discarded) and

$$r = R_2 - xx_2/R_2 - yy_2/R_2 + \left(x^2 + y^2\right)/(2R_2) \tag{4.12}$$

in which $R_2^2 = x_2^2 + y_2^2$ is the square of the distance between the land and the focal point. Since focusing occurs in the Fresnel diffraction region, we have retained second-order terms in x and y in the expression given by (4.12). The total distance between the source and the focal point is

$$R + r = -x\left(\frac{x_1}{R_1} + \frac{x_2}{R_2}\right) - y\left(\frac{y_1}{R_1} + \frac{y_2}{R_2}\right) + \frac{1}{2R_2}\left(x^2 + y^2\right) + R_1 + R_2 \quad (4.13)$$

and the amplitude at the focus is given by

$$E\left(x_2, y_2\right) = V \sum_{l=1}^{L} \exp\left\{i\frac{2\pi}{\lambda}\left[-y\left(\frac{y_1}{R_1} + \frac{y_2}{R_2}\right) + \frac{y^2}{2R_2}\right]\right\}$$
$$\times \int_{S_l} \exp\left\{i\frac{2\pi}{\lambda}\left[\frac{x^2}{2R_2} - x\left(\frac{x_1}{R_1} + \frac{x_2}{R_2}\right)\right]\right\} dx. \quad (4.14)$$

We thus see that the separation of variables can be carried out by examining the diffraction process in the kinematic approximation. The integral describes focusing by the zone plate in the Fresnel diffraction region. The integral is evaluated between limits corresponding to the formulas given by (4.4)–(4.6) for the ideal multilayer zone plate. The expression after the summation sign describes the selective properties of the multilayer Bragg–Fresnel layers:

$$G(y) = \sum_{l=1}^{L} \exp\left\{i\frac{2\pi}{\lambda}\left[-y\left(\frac{y_1}{R_1} + \frac{y_2}{R_2}\right) + \frac{y^2}{2R_2}\right]\right\}. \quad (4.15)$$

Substituting the corresponding expressions for the terms in (4.19), i.e., $y = ld$, $y_1/R_1 = \sin\theta_0$, $y_2/R_2 = \sin\theta_1$ we obtain

$$G(\theta_1) = \sum_{l=1}^{L} \exp\left\{i\frac{2\pi}{\lambda}\left[\frac{l^2 d^2}{2R_2} - ld\left(\sin\theta_0 + \sin\theta_1\right)\right]\right\}. \quad (4.16)$$

Let

$$P_l = \frac{2\pi}{\lambda}\left[\frac{i^2 d^2}{2R_2} - 2id\sin\theta_0\right]. \quad (4.17)$$

This takes account of the fact that the reflection maximum lies in the direction of the specular angle for $y_1/R_1 = y_2/R_2$. The selectivity of reflection can then be written in the form

$$|G(\theta_1)|^2 = \left[\left(\sum_{l=1}^{L} \sin P_l\right)^2 + \left(\sum_{l=1}^{L} \cos P_l\right)^2\right] \Big/ L^2. \quad (4.18)$$

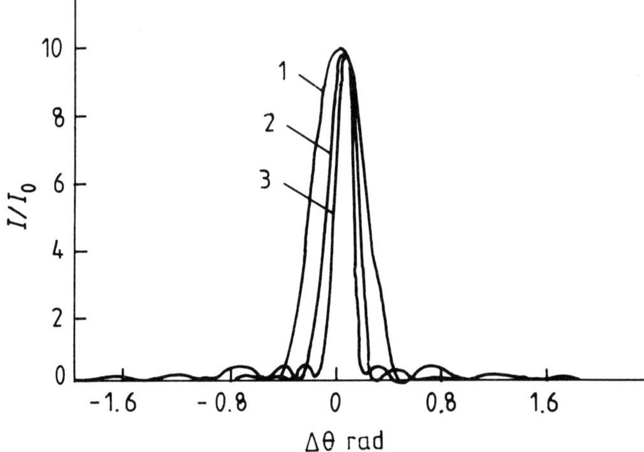

Figure 4.6 Angular selectivity for a Bragg–Fresnel lens for $L = 10$ (1), 20 (2) and 30 (3).

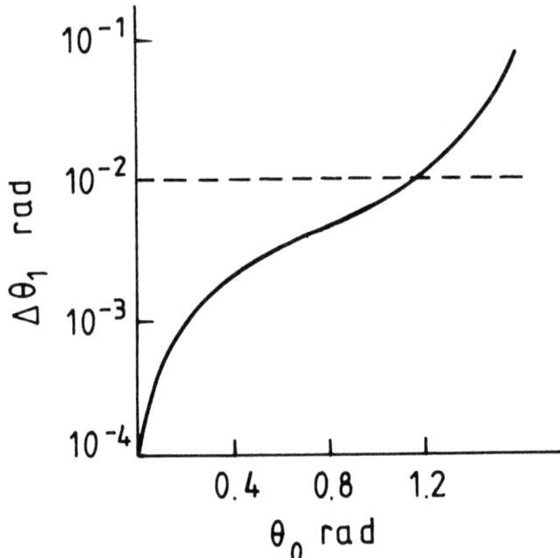

Figure 4.7 Angular width of the BFL angular selectivity curve for $L = 100$ in the kinematic approximation.

Figure 4.6 shows the curves describing the angular selectivity of the BFL for different numbers of layers. The presence of the quadratic Fresnel term in (4.17) produces a shift of the distribution maximum from the exact Bragg angle. The quadratic term in (4.21) can be neglected when the calculate the width of the curve, so that shape of the curve is described by

$$|G(\theta_1)|^2 = \frac{\sin^2(2\pi L d \sin\theta_1/\lambda)}{L^2 \sin^2(2\pi d \sin\theta_1/\lambda)}. \tag{4.19}$$

If we suppose that $\Delta\theta_1 \ll 1$ is a small departure from the exact Bragg angle, then (4.19) shows that the first zero of the curve is observed for

$$\Delta\theta_1 = \sqrt{\cot^2\theta_0 + 2/L} - \cot\theta_0 \qquad (4.20)$$

where θ_0 is the Bragg angle.

Figure 4.7 shows $\Delta\theta_1$ as a function of the Bragg angle for a multilayer structure containing $L = 100$ layers. Although the expressions given by (4.19) and (4.20) are only very approximate, they do provide good estimates for the properties of multilayer BFLs. The restricted spectrum of spatial frequencies transmitted by the BFL means that, for small angles of incidence, the BFL acts mostly as a focusing element with a very small field of view. The quantity $\Delta\theta_1$ is effectively the angular field of view of the BFL element. For example, the focal-spot size for an extended source is determined exclusively by the quantity

$$\delta_x = \Delta\theta_1 F \qquad (4.21)$$

where F is the focal length of the BFL. The selected properties of the BFL with respect to the wavelength of the incident radiation are similar to those encountered in pure Bragg diffraction by a multilayer structure:

$$\lambda/\Delta\lambda = N. \qquad (4.22)$$

Considerable efforts are being devoted to the development of Bragg–Fresnel lenses based on perfect silicone and germanium crystals for the hard X-ray region [158,159]. The first report of the modulation of the wave incident on the crystal by the periodic structure of the surface is given in [158]. The description of diffraction by substructures is more complicated and requires the use of the dynamic theory of X-ray diffraction.

4.2.2 Fabrication of Bragg–Fresnel optical elements

The structure of the BFL elements calculated from (4.4)–(4.6) can be fabricated in practice in a number of different ways.

The amplitude BFL can be produced in two ways, namely, by masking the surface of a multilayer mirror with a highly absorbing material in accordance with the shape of the Fresnel zones or by etching corresponding segments of the multilayer mirror (figure 4.8). The amplitude BFL has a low efficiency and by analogy with transparent zone plates, at least 50% of the incident radiation is absorbed by the substrate.

Phase Bragg–Fresnel lenses can also be produced in a number of ways. The simplest way is to deposit a binary phase-shifting layer producing a phase difference by $\lambda/4$ on the surface of a multilayer mirror. The more complicated way is to deposit a special kinoform profile in the form of a phase-shifting material on the surface of the multilayer mirror. Finally, the most promising method is to deposit a multilayer mirror on a pre-profiled substrate or to create the profile directly in the multilayer structure (figure 4.9). The total efficiency of the Bragg–Fresnel lens can then reach 20–70% depending on the wavelength.

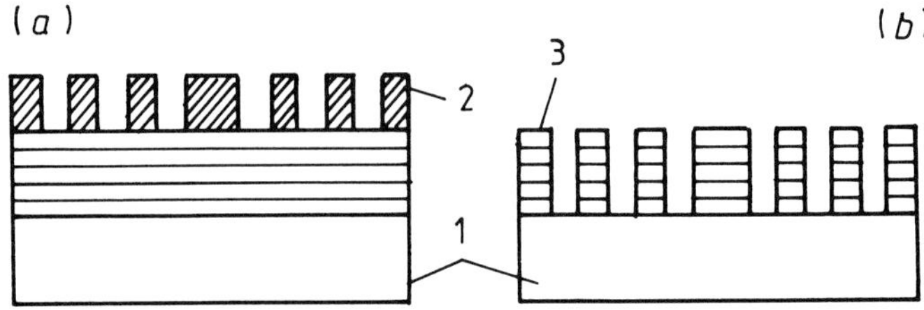

Figure 4.8 Amplitude-phase Bragg–Fresnel elements produced by forming an amplitude phase mask on the surface of the mirror (*a*) and by etching out selected portions of the multilayer structure (*b*): (1) substrate, (2) mask made from absorbing or phase medium, (3) multilayer mirror.

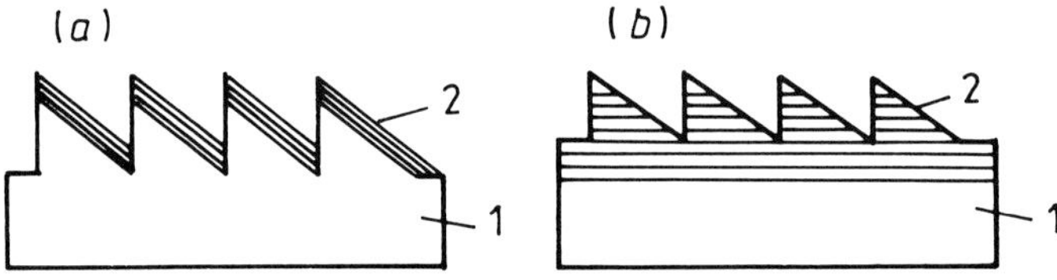

Figure 4.9 Production of kinoform Bragg–Fresnel elements by successive deposition of layers on a polished surface (*a*) and by profiling a multilayer structure (*b*) : (1) substrate, (2) multilayer structure.

The production of a kinoform element on a profile substrate is a very difficult problem. This is so because the shape of the surface must be accurate to within a few ångströms, since otherwise the phase relationships will be violated. This means that the only possibility, at least at present, is to produce a binary profile on the substrate. A kinoform element produced within a multilayer structure is a more realistic proposition (figure 4.9(*b*)) since this has to be done to within the extinction length, 0.1–0.3 μm; actually there is a phased difference because of the different optical part length in the multilayer structure and in vacuum. This difference is equal to π radians for a step height equal to the extinction length. We note that this method can be used to produce diffraction gratings (echelettes) of very high resolving power (10^5–10^6) [160].

Fabrication of linear BFL. The first experimental implementation of X-ray focusing by BFL on a multilayer mirror is reported in [161] and the corresponding result for single-crystal silicon is described in [159].

The substrates for the Bragg–Fresnel lenses were multilayer mirrors produced by the laser deposition method at Institute of Applied Physics of the Russian Academy of Science and by the ion-beam deposition method at the Institute of Microelectronics Technology of the Russian Academy of Sciences. The

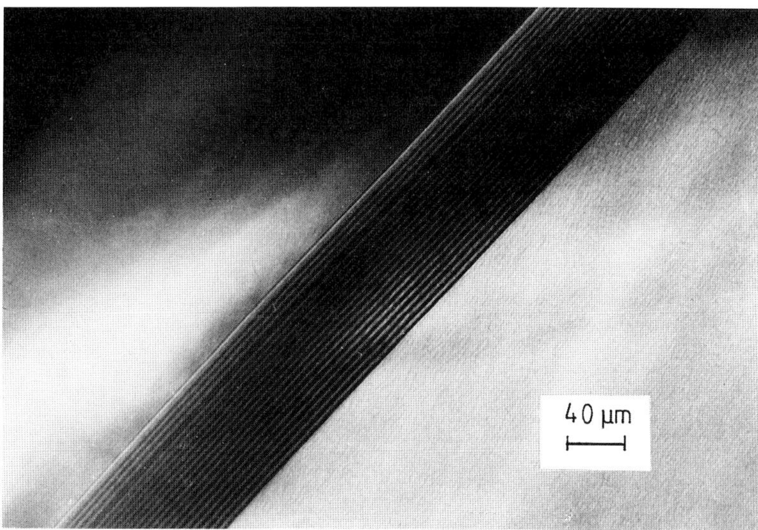

Figure 4.10 Micrograph showing the cross section of a multilayer mirror (tungsten/carbon) fabricated at the Institute of Microelectronic Technology of the Russian Academy of Sciences. Period of the structure is 4 nm.

mirrors incorporate alternating layers of nickel and carbon and also tungsten and carbon. The period of the structure is between 4.5 and 3.7 nm. Figure 4.10 shows a cross section of the multilayer structure produced in a transmission electron microscope.

The first experimental results were obtained with linear zone plates, [161,162]. Figure 4.11 illustrates the sequence of operations used to fabricate zone plates with segments of the multilayer mirror removed (a) and with a phase-shifting layer deposited on the mirror surface (b). An electron-beam lithography mask was drawn on the surface of a multilayer mirror coated in a centrifuge with an electron resist. Prototype Fresnel lenses were then produced by two different methods. The profiled amplitude BFL were etched through the resist mask in a beam of 1 keV argon ions for 20 minutes [160,164]. During this process, the portion of the mirror that was not protected by the resist was removed from the substrate. On other samples, a thin layer of gold (30 nm) was deposited within the resist windows by an electrolytic method. When the resist was removed, the surface of the mirror was masked over particular areas by the gold layer approxiamtely 30 nm thick.

The arrangement illustrated in figure 4.12 was used in an experimental verification of the focusing properties of the BFL fabricated in this way. The source of radiation was an X-ray tube with a Fe anode. The radiation wavelength was 0.196 nm and the corresponding Bragg angle for a structure with a period of 4.8 nm was 1.15 degrees (2.45 mrad). The size of the central zone of the zone

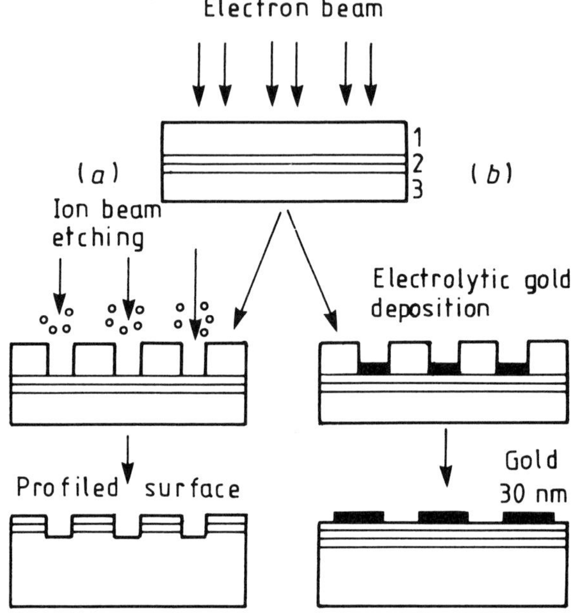

Figure 4.11   A diagram illustrating the sequence of operations used to fabricate (*a*) and phase (*b*) Bragg–Fresnel lenses: (1) electron resist, (2) multilayer mirror, (3) substrate.

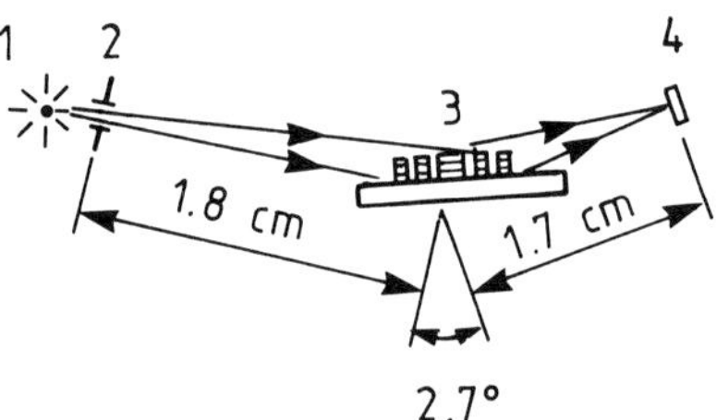

Figure 4.12   Experimental arrangement used to test linear BFL with a short-wave source of X-rays [161]: (1) X-ray tube, (2) entrance slit, (3) linear BFL, (4) recording device.

plate was $90\,\mu$m; the projection onto the direction of the incident beam was roughly was $1.8\,\mu$m. Consequently, the focal length of this BFL was 1.7 cm. The minimum size of a zone on this structure was $9\,\mu$m and the corresponding projection onto the direction of the incident beam was $0.2\,\mu$m. The X-ray beam from the tube was collimated by a two-slit collimator with a slit width of $50\,\mu$m. The intensity distribution in the focal plane was recorded on high-resolution photographic plates MIKRAT VRE. Figure 4.13 shows the density distribution recorded in the focal plane. The diameter of the focal spot ($6\,\mu$m) corresponds to a fivefold reduction in the entrance-slit image. Photographs recorded at different

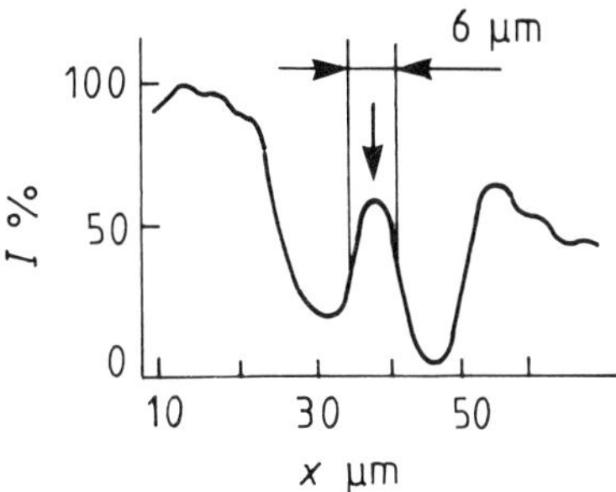

Figure 4.13 Intensity distribution in the plane of the image produced by the linear BFL. Arrow shows the position of the focal spot.

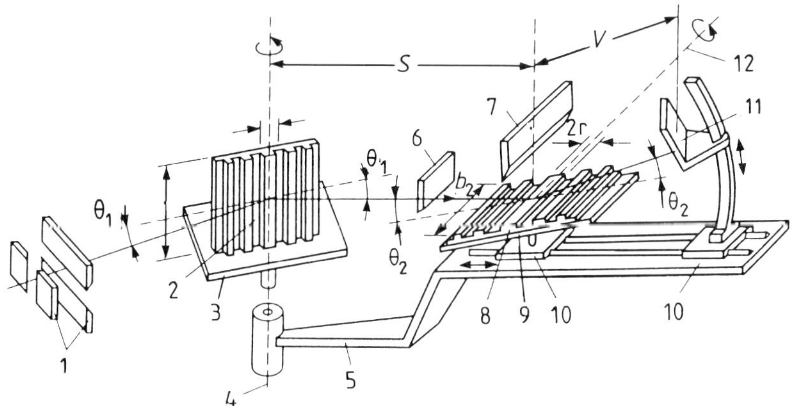

Figure 4.14 The Kirkpatrick–Baez microscope with two zone plates: (1) entrance slits, (2) BFL-1, (3) stage for BFL-1, (4) rotation axis of BFL-1, (5) rotating stage of the goniometer and BFL-2, (6, 7) limiting zero-order slits, (8) BFL-2, (9) stage for BFL-2, (10) guiding rods for BFL-2 and the detector, (11) detector, (12) rotation axis of BFL-2.

distances from the image plane show that the BFL has the same properties as an ordinary lens.

Kirkpatrick–Baez microscope using Bragg–Fresnel lenses. The production of high-resolution, two-dimensional image-transfer systems working at short wavelengths of approximately 0.1 nm is a relatively complicated problem. Grazing angles of incidence can be used to achieve in the one-dimensional case a projection of zones onto the direction of the incident beam with a 'compression factor' of up to 100. In the case of conocal diffraction (along the axis of the incident beam), the zone dimensions remain very small and are inaccessible to fabrication by existing techniques. However, two-dimensional focusing can

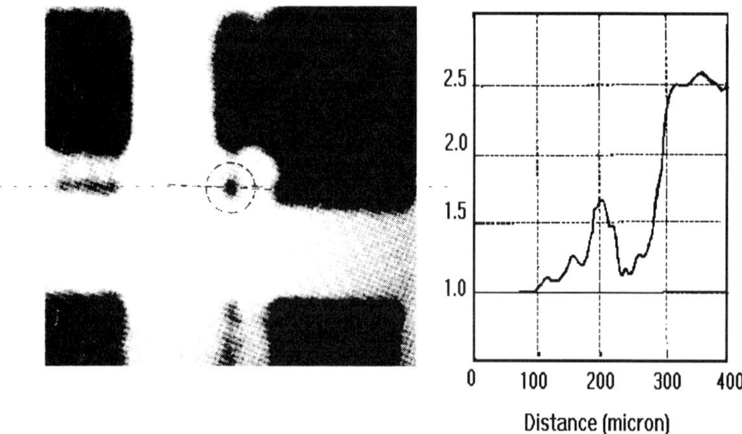

Figure 4.15   Micrograph and density distribution obtained for the image focused by the Kirkpatrick–Baez microscope using linear multilayer BFL.

probably be achieved by using the Kirkpatrick–Baez scheme, i.e., separating focusing along two perpendicular directions.

Figure 4.14 illustrates this type of microscope with two Bragg–Fresnel multilayer lenses. The BFLs used in this experiment were designed to have different focal lengths and contain a large apodized region (1.5 mm) used to suppress the influence of zero-order diffraction. The minimum zone size is 5 μm so that photolithography can be used to fabricate such structures. The expected resolving power of the system is 0.1 μm. The Kirkpatrick–Baez diffraction microscope was tested experimentally using an X-ray tube with radiation of wavelength 0.154 nm. Figure 4.15 shows the image recorded in the focal plane of the microscope. The size of the focal spot was 16 μm and dependent on the geometry of the system.

Ellipsoidal Bragg–Fresnel lenses. The structure of a Bragg–Fresnel optical element producing a circular focal spot was discussed in detail in section 4.2.1. It is described by the parametric equations given by (4.4)–(4.6). However, the topology is more readily produced in the electron-beam lithographic system by using Cartesian coordinates and the spline approximation. The control system and its programmed implementation are discussed in greater detail below.

Transforming (4.4)–(4.6) we obtain the equation of the ellipsoidal section in cartesian and ellipsoidal coordinates:

$$z^2 = \left(\frac{1}{\xi_n^2} - 1 - k^2 \right) \left(\frac{x}{p} + x_0 + \frac{kb_1\xi_n^2}{1 - p^2\xi_n^2} \right)^2 - b_l^2$$
$$+ a^2 \left(\xi_n^2 - 1 \right) - \frac{k^2 b_l^2 \xi_n^2}{1 - p^2\xi_n^2} \tag{4.23}$$

where $x_0 = (ak/p^2)(a^2 + b_l^2)$, $p = \sqrt{k^2 + 1}$.

If we now expand the ellipsoidal coordinate ξ_n in accordance with (4.12), we obtain

$$\xi_n = \xi_0 + n\lambda/(4a). \tag{4.24}$$

Consequently, neglecting second-order terms in λ, we obtain

$$\xi_n^2 = \xi_0^2 + \xi_0 n\lambda/(2a). \tag{4.25}$$

Equation (4.23) gives the exact solution for the coordinate of zones in three-dimensional states. Approximate working formulas can be obtained (4.25). Substituting this equation in (4.23) and rearranging we can readily show that

$$\left(x + \frac{ka^2 cn\lambda/2}{b_l^3 + pb_l cn\lambda/2}\right)^2 + z^2 \left(\frac{c^2 + pcn\lambda/2}{b_l^2 + pcn\lambda/2}\right)$$
$$= \frac{(c^2 + pcn\lambda/2)(b_l^2 - a^2 k^2)cn\lambda/2}{\left(b_l^2 + pcn\lambda/2\right)^2} \tag{4.26}$$

where $c = \sqrt{a^2 + b_l^2}$.

If we now assume that

$$a \approx b_l \gg 0.5\,n\lambda \tag{4.27}$$

we find that (4.30) assumes the simpler form

$$\left(x_n + \frac{k^2 a^2 c\lambda n}{2b_l^3}\right)^2 + \frac{c^2}{b_l^2}z^2 = \frac{c^2}{b_l^2}Fn\lambda \tag{4.28}$$

where F is the focal length of the Bragg–Fresnel lens. In practice, the focal length is calculated from the distance between the source and the focal point and the necessary magnification as follows. If the distance between the source, (object) and the centre of the BFL is denoted by R_1 and the distance to the focal point by R_2 then the known properties of ellipsoidal coordinates lead to

$$R_1 + R_2 = 2a\xi_0 \qquad R_1 - R_2 = 2a\eta_0. \tag{4.29}$$

The classical lens magnification formula

$$M = R_1/R_2 \tag{4.30}$$

thus becomes

$$M = (b_0 + ak)/(b_0 - ak) \tag{4.31}$$

where

$$\eta_0 = ak\xi_0/b_0. \tag{4.32}$$

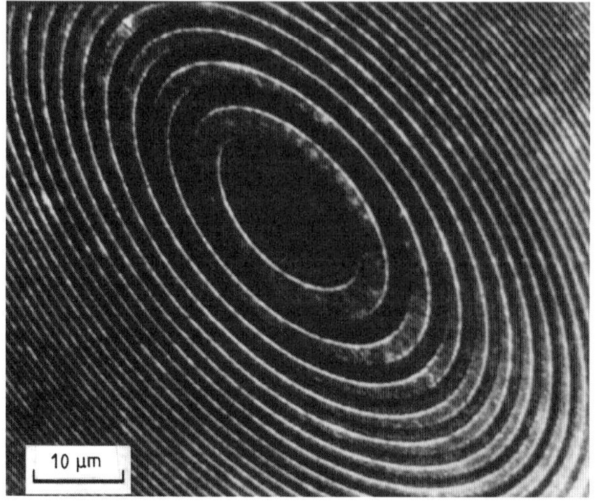

Figure 4.16   Ellipsodial Bragg–Fresnel lens [153].

From (4.31) the expression for the slope k becomes

$$k = b_0 (M - 1) / [a (M + 1)].$$
(4.33)

Similarly, from (4.29) and the lens formula, we find that the focal length of the BFL is

$$F = \frac{\sqrt{a^2 + b_0^2} \left(b_0^2 - a^2 k^2\right)}{2 b_0^2 \sqrt{k^2 + 1}}.$$
(4.34)

The parameters b_0 and a are related to the period of the multilayer structure and the wavelength of incident radiation by the simple expression

$$b_0 = a\lambda\sqrt{4d^2 - \lambda^2} = a \tan\theta_B.$$
(4.35)

Let us now consider the special case of a Bragg–Fresnel lens in the form of a polarizer in which the incident radiation is reflected at $90°$. When the angle of incidence is $45°$, we find from (4.35) that $b_0 = a$. Equation (4.28) which describes the structure of the zone on the surface of the BFL now assumes the form

$$\left(x + kn\lambda/\sqrt{2}\right)^2 + 2z^2 = 2Fn\lambda$$
(4.36)

and describes ellipsoids that are elongated along the axis with a variable shift relative to the origin of coordinates by the amount $kn\lambda\sqrt{2}$. If the angle of incidence is $45°$ we have $k = (M - 1)/(M + 1)$ and

$$F = a \left(1 - k^2\right) / \sqrt{2 \left(k^2 + 1\right)}$$
(4.37)

for the surface of the multilayer structure.

Examples of ellipsoidal Bragg–Fresnel lenses based on multilayer structures are reported [153,164,165]. Ni/C with $d = 3.2$ nm was used as the starting point. The mirrors were fabricated at the Institute of Applied Physics of the Russian Academy of Sciences by the laser deposition method [145]. The maximum reflection coefficient at the $K\alpha$ line of carbon (4.7 nm) was found to be 8–11%. The structure of the ellipsoidal BFL fabricated in accordance with (4.37) is shown in figure 4.16. Its principal parameters were as follows:

external dimensions	$360 \times 507\ \mu$m
total number of zones	360
focal length at the wavelength of 4.5 nm	20 mm
width of outermost zone	$0.25\ \mu$m.

The BFL shown in figure 4.16 is an amplitude device and was fabricated by etching out a multilayer structure (figure 4.10(b)).

The focusing properties of ellipsodal BFLs were investigated in synchrotron radiation (SR) from the electron storage ring Siberia-1 at the Kurchatov Institute of Atomic Energy. Figure 4.17 shows a schematic diagram of the experimental station mounted in the radiation channel of the storage ring. The synchrotron-radiation beam had a broad 'white' spectrum and was intercepted by a stop before preceeding to the BFL mounted on a special stage at 45° to the beam axis. The position of the BFL could be adjusted visually in the light of the visible component of the incident SR beam. The radiation was recorded on a photographic plate protected from visible radiation by a polyimide filter and an aluminum film 50 nm thick. The radiation emerging from the bending magnet of the storage ring was used in the experiment.

Figure 4.18 shows the intensity distribution in the image produced in the focal plane of the BFL [166]. This experiment yielded images of the 2.8×5 mm source of radiation reduced by a factor of 350.

It was found in these experiments that the shape of the focal spot was very dependent on the angle of rotation of the axis of the ellipse relative to the beam axis.

4.2.3 Aberrations of Bragg–Fresnel lenses

The properties of the Bragg–Fresnel lenses as imaging systems are determined by their aberrations. These abberrations can be analysed by expanding the optical difference between the rays as described in chapter 2 for the aberrations of a zone plate.

First, we note that if the BFL is fabricated in accordance with (4.4)–(4.6) and is a complete representation of the spatial distribution of the three-dimensional Fresnel zones, then there is no spherical aberration. This aberration can only appear if there is a discrepancy between the actual three-dimensional profile of

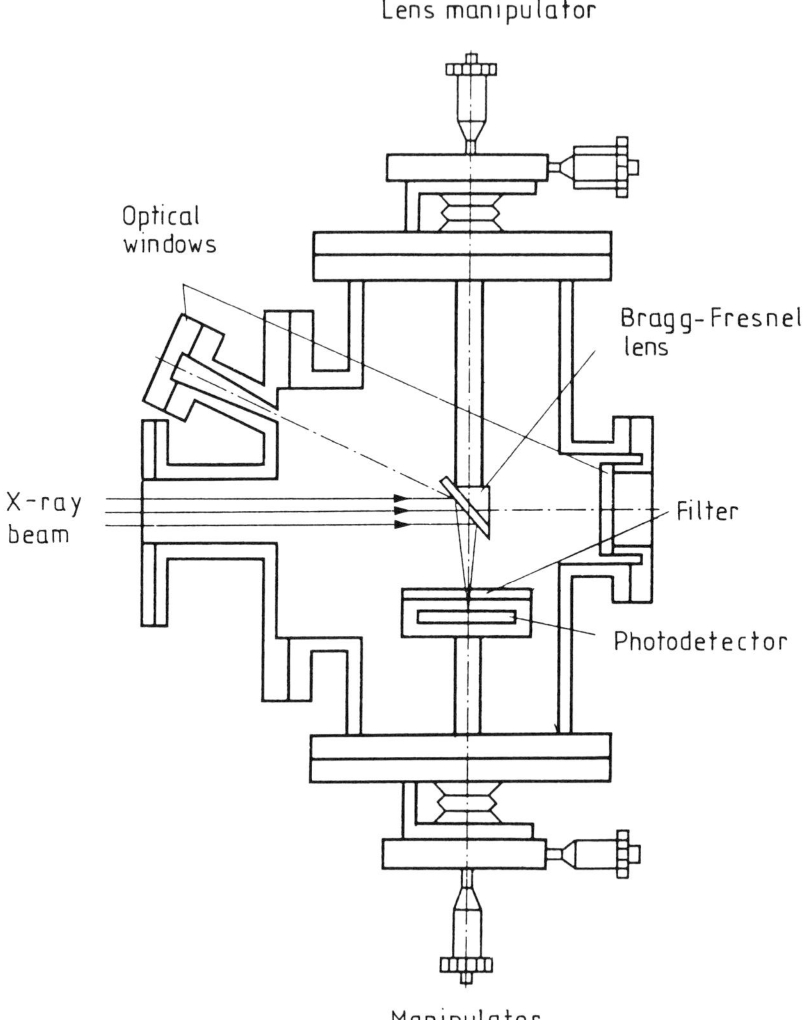

Figure 4.17 Experimental station for testing ellipsoidal BFLS on the Siberia-1 electron storage ring [166].

the zones and the calculated profile. In particular, the fabrication of a zone profile that varies with depth is still an unresolved technological problem. From the practical point of view, it is therefore interesting to analyse the possibilities of multilayer BFL with a rectangular zone profile.

Spherical aberrations of BFL with rectangular zone profile. In accordance with (4.35) to achieve focusing at the same point, the structure of each layer must have its own focal length F_l. Let us define the layer parameter by

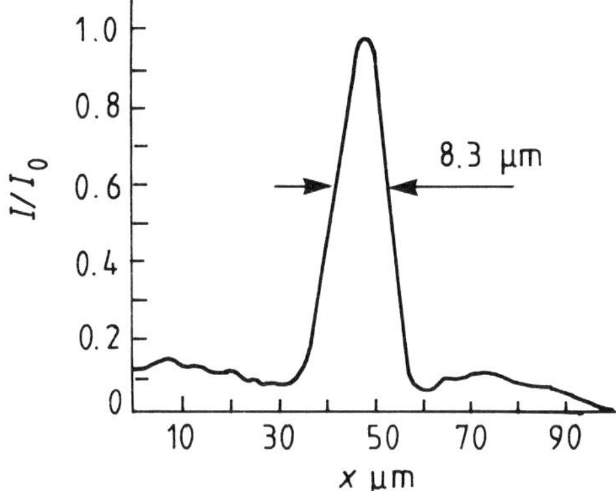

Figure 4.18 Intensity distribution recorded in the focal plane of the ellipsoidal BFL [166].

$$b_l = g_l a k (M - 1)/(M + 1) \tag{4.38}$$

where $g_l = 1 + ldp/a$. We can then write

$$F_l = a \left(g_l^2 m^2 - 1\right) \left(1 + g_l^2 k^2 m^2\right)^{1/2} / \left(2 p g_l^2 m^2\right) \tag{4.39}$$

where F_l is the focal length of layer l in the multilayer structure and $m = (M - 1)/(M + 1)$. The expression for the width of the nth zone is

$$\Delta r_{ni} = 0.5 \left(F g_{nl} \lambda / n\right)^{1/2} . \tag{4.40}$$

If the BFL on the multilayer mirror has a rectangular profile with vertical walls, the calculated Fresnel structure within the mirror layers is not the same as the fabricated profile. The error in the difference between the radii of the lower zones and the ideal value can be calculated by demanding that

$$r_{n0} - r_{nl} < \lambda/4 \tag{4.41}$$

where r_{n0} and r_{nl} represent the zone radii on the surface and within the multilayer structure, respectively. The condition for the absence of large aberrations that result from this discrepancy between the dimensions of the zones can be taken to be the Rayleigh criterion, i.e.,

$$0.5 \Delta r_n > r_{n0} - r_{nl}. \tag{4.42}$$

Using the approximate expression for the width of the nth zone, we then have

$$N \leqslant 0.25 \sqrt{F_0} \left(\sqrt{F_0} - \sqrt{F_l}\right)^{-1} . \tag{4.43}$$

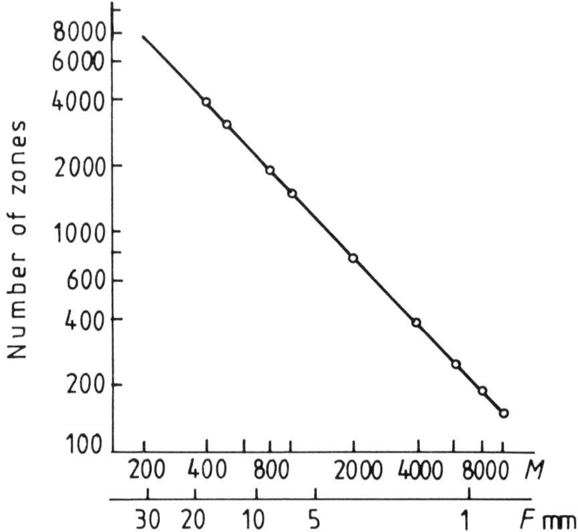

Figure 4.19 Limiting number of zones in a Bragg–Fresnel lens as a function of the magnification factor and the focal length. The number of layers in the multilayer structure is 100, its period is 3.2 nm, the grazing angle is 45° and $2a = 700$ cm.

Estimates made for the actual experimental conditions have led to the following results. When the separation between the source of radiation (object) and the focal plane is $2a = 700$ cm, the reduction factor is 350, the period of the mirror is $d = 3.2$ nm (reflection at 45°), and the number of layers is 100, the calculated value of N is 4400.

Figure 4.19 shows the calculated maximum number of zones as a function of the reduction factor M and the focal length F of the lens for a particular experimental realization. When $M \approx 1000$ the working number of zones producing aberration-free images reaches 1500.

Off-axis aberrations of the Bragg–Fresnel lens. The off-axis aberrations of the BFL can be analysed on the basis of the data obtained for spherical aberration. It follows from the graph in figure 4.19 that, for a BFL with a rectangular zone profile and a focal length in excess of 5 nm, an aberration-free image is formed by about 1000 zones (lancing angle of 45°). Microzone plates fabricated by electron-beam lithography have at best a few hundred zones. This means that when we analyse the aberration properties of the multilayer BFL in which the number of zones is of the order of 100 we can confine our attention to the optical path difference associated with diffraction on the surface of the structure. Using the notation of figure 4.20, the optical path difference in the case of a plane wave incident on the BFL can be written in the form of

$$\Delta L = \mathrm{BC'} - \mathrm{A'C'}. \tag{4.44}$$

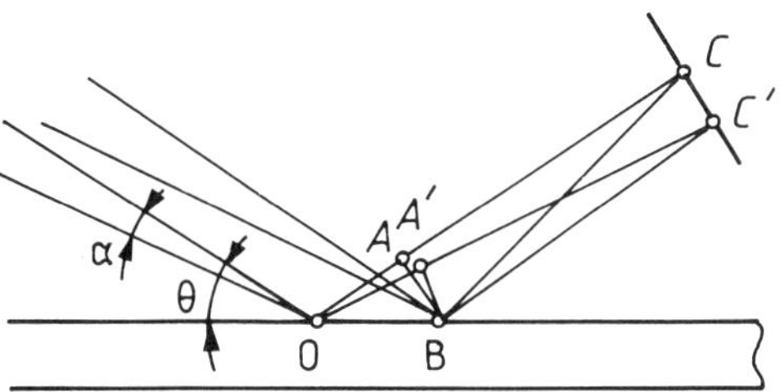

Figure 4.20   Calculation of the optical path difference for a Bragg–Fresnel lens.

We now introduce the following notation: $r = r_n \sin\theta$ is the effective zone size, θ is the glancing angle of the incident beam, $f = r_n \cos\theta$ is the projection of the BFL on the beam axis and $F_1 = F + r_n \cos\theta$ is the effective focal length. This gives

$$\mathrm{BC'} = F_1 + \frac{r^2}{2F_1} - r\alpha + \frac{\alpha^2}{2}F_1 + \frac{r^3\alpha}{2F_1^2} - \frac{3\alpha^2 r^2}{4F_1} \tag{4.45}$$

$$\mathrm{A'C'} = F_1 + F\frac{\alpha^2}{2} - r\alpha. \tag{4.46}$$

Substituting the last two expressions into (4.48) we find that

$$\Delta L = \frac{r^2}{2F_1} + \alpha\left(\frac{r^3 F}{2F_1^3} + \frac{rf}{F_1}\right) - \alpha^2\left(\frac{Ff}{2F_1} + \frac{3r^2 F^2}{4F_1^3}\right). \tag{4.47}$$

The first term (in this expression) represents the coma whereas the second term represents astigmatism and field curvature. Comparison of (4.47) with the analogous expression for axial zone plates shows that large aberration terms have appeared and are proportional to the cosine of the glancing angle:

$$\alpha\left[r^3 F/\left(2F_1^3\right) + rf/F_1\right] \qquad \text{coma}$$
$$\alpha^2\left[3r^2 F^2/\left(4F_1^3\right) + Ff/\left(2F_1\right)\right] \quad \text{astigmatism and field curvature.}$$

The focusing conditions are met if $r^2 = nF_1 = \lambda$ and, hence,

$$r_n = \frac{\lambda n \cos\theta}{2\sin^2\theta} \pm \sqrt{\frac{n^2\lambda^2 \cos^2\theta}{4\sin^2\theta} + \frac{nF\lambda}{\sin^2\theta}}. \tag{4.48}$$

When the beam is incident normally on the surface of the BFL, we have $f = 0$ and the aberrations are completely analogous to those of a transparent zone plate. As the beam departs from the normal, the size of the aberration-free field of view falls sharply.

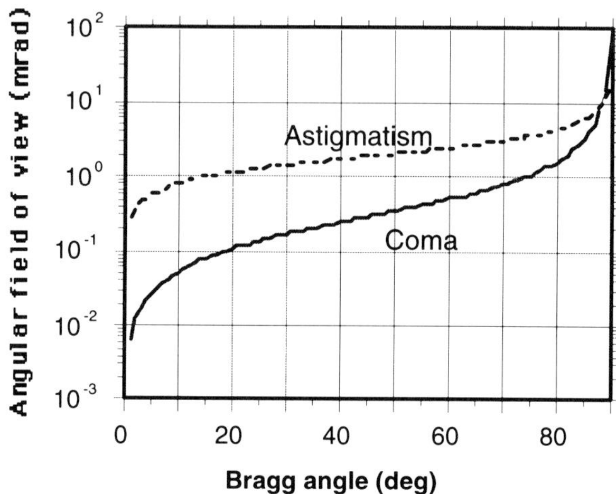

Figure 4.21 Maximum field angle of the BFL as a function of the Bragg angle with allowance for coma and astigmatism.

Below, we present an example of a calculation of the aberration properties of the BFL for wavelengths of 5.9 nm and focal length of 1.5 cm. First, we must estimate which aberrations are the dominant ones. As shown in chapter 2, when $\theta = 90°$ and the number of zones is small, the dominant contribution is provided by astigmatism. As the incident beam departs from the normal, all the off-axis aberrations grow rapidly and the maximum field angles corresponding to the aberration-free field are given by the following expressions:

$$\alpha_{\text{coma}} = \frac{\Delta r_{\min}^2}{\lambda F}\left[\left(\frac{\lambda}{4\Delta r_{\min}\left(1 + \frac{\lambda}{2\Delta r_{\min}}\cot\theta_B\right)^3}\right) + \left(\frac{\cot\theta_B}{1 + \frac{\lambda}{2\Delta r_{\min}}\cot\theta_B}\right)\right]^{-1}$$

$$(4.49a)$$

$$\alpha_{\text{ast.,field curv.}}$$

$$= \sqrt{\frac{\Delta r_{\min}}{F}\left[\left(\frac{3\lambda}{4\,\Delta r_{\min}\left(1 + \frac{\lambda}{2\,\Delta r_{\min}}\cot\theta_B\right)^3}\right) + \left(\frac{\cot\theta_B}{1 + \frac{\lambda}{2\,\Delta r_{\min}}\cot\theta_B}\right)\right]^{-1}}$$

$$(4.49b)$$

where $\Delta r_{\min} = 0.5(\lambda f/N)^{1/2}$ is the minimum zone size and θ_B is the Bragg angle.

Figure 4.21 shows the maximum angular size of the aberration-free image field for 100 Fresnel zones ($N = 100$, $F = 1.5$ cm) and different beam grazing angles.

When the grazing angle is less than $90°$, the coma is the dominant aberration for

$$\tan\theta \leqslant 8N^{3/2}\sqrt{\lambda/f} \tag{4.50}$$

where $r \approx (NF\lambda)^{1/2}/\sin\theta$ which follows from (4.48). The field angle of the aberration-free image can therefore be estimated from the relation

$$\alpha \approx \tan\theta/(2N) \qquad N \geqslant \tan\theta\sqrt{F/\lambda}/8. \tag{4.51}$$

Thus, for $N = 100$, $\tan\theta = 1$, $F = 1.5\,\mathrm{cm}$ and $\alpha = 1.25 \times 10^{-3}$, the size of the image field is $38\,\mu\mathrm{m}$. We note that this value falls on the curve of the multilayer structure. The small field of view of the Bragg–Fresnel lens is obviously a restriction on their amplification on imaging systems. The BFL is most suitable for scanning systems.

4.3   X-RAY MODULATORS

The X-ray refractive indices and absorption coefficients of different media are not very different from unity and are not very sensitive to physical effects occuring in these media even when the wavelength of the incident radiation lies near the absorption edge of an element. This means that conventional modulation techniques used for optical radiation cannot be extended to the X ray range. X-ray modulators must rely on coherent Bragg scattering and are based on the following physical phenomena.

(1) For the amplitude modulation of an incident wave, the crystal or multilayer mirror must be taken out of its reflecting position. For example, for a multilayer mirror, the layer oscillation amplitude should be

$$\tilde{h} = \Lambda \tan\theta_0/(2\pi L) \tag{4.52}$$

where Λ is the wavelength of the oscillations on the surface. For example, when $\theta_0 = 45°$, $L = 100$ and $\Lambda = 10\,\mu\mathrm{m}$, we have $h = 16\,\mathrm{nm}$ (see (4.70)).

(2) For phase modulation, the grating must be shifted by half the plane separation. For the usual values of the plane separation, the magnitude of $\tilde{h}$ of multilayer structures is of the order 2.5–5 nm. This can be used to reduce the deformation of the grating necessary for modulation by two orders of magnitude as compared with (4.52).

(3) The deformation of the grating necessary to produce amplitude and phase modulation can probably be reduced still further by using diffraction by complex multilayer structures with a variable period, structure-forbidden reflections, and strong excitation. If we use radiation whose wavelength lies near the absorption edge of one of the elements present in the multilayer mirror, modulation can be produced even without varying the plane separation if the absorption edge can be shifted by some external means.

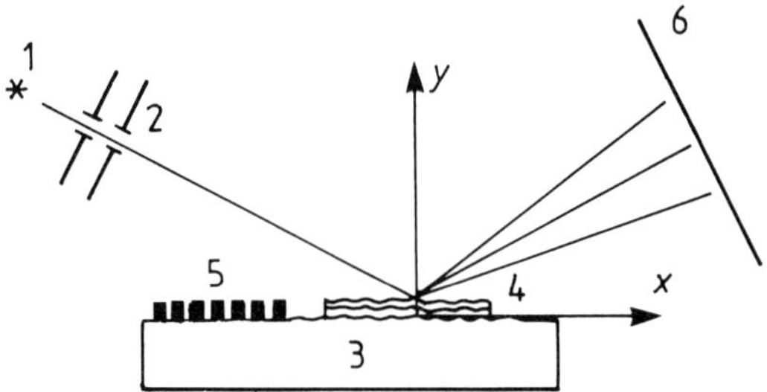

Figure 4.22 Arrangement used to observe X-ray diffraction by a multilayer mirror modulated by a surface acoustic wave: (1) source of radiation (X-ray tube), (2) collimator, (3) quartz substrate, (4) multilayer mirror, (5) source of surface acoustic wave, (6) photographic plate.

The first X-ray modulation experiments were performed using crystals and short wavelengths ($\sim 0.2\,\text{nm}$). Several workers have investigated the effect of bulk acoustic waves [167–169] and surface acoustic waves [170,171] on the working crystal. In [171], X-ray amplitude modulation as described by (4.52) was realized in a perfect crystal using a bulk acoustic wave. In [172], phase modulation was achieved by displacing the reflecting planes at different points of the crystal.

In the soft X-ray range, the diffraction of X-rays by a multilayer structure modulated by a surface acoustic wave was reported in [173].

The physics of the diffraction of X-rays by the surface of a multilayer mirror modulated by a surface acoustic wave can be described qualitatively in the kinematic approximation by analogy with (4.12) [174]. Consider a multilayer interference mirror as a set of L parallel partially reflecting planes with plane separation d (figure 4.22). The Y axis points along the normal to the multilayer mirror. For the lth layer in the multilayer mirror we then have the amplitude

$$y = ld + \tilde{h}\cos(Kx) \qquad K = 2\pi/\Lambda. \tag{4.53}$$

Suppose that an X-ray wave falls on the multilayer mirror from a distant point x_1, y_1. We shall determine the field at a point x_2, y_2 in the far-field region. In the kinematic approximation, in which attenuation by diffraction of incident radiation and secondary scattering of diffracted waves can be neglected, the field at the point of observation is given by (4.19). To calculate the relative intensity of diffraction orders in the far-field region (Fraunhofer) we shall use the approximation

$$R = R_1 + xx_1/R_1 + yy_1/R_1$$
$$r = R_2 + xx_2/R_2 + yy_2/R_2 \tag{4.54}$$

where, as in section 4.2.1, R_1 and R_2 are the distances of the points (x_1, y_1) and (x_2, y_2) from the origin. The field at (x_2, y_2)

$$E(x_2, y_2) = V \sum_{l=1}^{L} \int_{S_l} \exp\left\{ i\frac{2\pi}{\lambda} \left[x\left(\frac{x_1}{R_1} + \frac{x_2}{R_2} \right) \right. \right.$$

$$\left. \left. + \left(ld + \tilde{h}\cos(Kx) \right)\left(\frac{y_1}{R_1} + \frac{y_2}{R_2} \right) \right] \right\}\, dx. \qquad (4.55)$$

We now expand the exponential in terms of the Bessel functions $J_m(\phi)$:

$$\exp\left[i\varphi\cos(\beta x) \right] = \sum_{m=-\infty}^{\infty} i m\, J_m(\varphi_1) \exp(im\beta x) \qquad (4.56)$$

so that the field amplitude at the point of observation (x_2, y_2) can be written in the form

$$E(x_2, y_2) = V \sum_{l=0}^{L} \sum_{m=-\infty}^{\infty} \exp\left[i\frac{2\pi}{\lambda} ld\left(\frac{y_1}{R_1} + \frac{y_2}{R_2} \right) \right] i^m J_m\left[k\tilde{h}\left(\frac{y_1}{R_1} + \frac{y_2}{R_2} \right) \right]$$

$$\times x \int_{S_l} \exp\left\{ ix\frac{2\pi}{\lambda}\left(\frac{x_1}{R_1} + \frac{x_2}{R_2} \right) + mK \right\}\, dx \qquad (4.57)$$

The integral in this expression is a δ-function and is non-zero only for values of the argument defined by

$$2\pi x_2/(\lambda R_2) = 2\pi x_1/(\lambda R_1) + mK. \qquad (4.58)$$

The direction cosines of the diffraction orders are given by $\cos\theta_m = \cos\theta_0 + m\lambda/\Lambda$. It follows that the field diffracted by the surface has special directions corresponding to diffraction orders m produced by the modulated multilayer structure. The signs of these angles corresponding to these diffraction orders are given by $\sin\theta_m = y_2^{(m)}/R_2$. The fields amplitudes corresponding to these diffraction orders are given by

$$E_m = V i^m J_m\left[\frac{2\pi}{\lambda}\tilde{h}\left(\frac{y_1}{R_1} + \frac{y_2^{(m)}}{R_2} \right) \right] g_m(L) \qquad (4.59a)$$

where

$$g_m(L) = \sum_{l=1}^{L} \exp\left[-\mu_l ld\left(\frac{y_1}{R_1} + \frac{y_2^{(m)}}{R_2} \right) \right] \exp\left[i\frac{2\pi}{\lambda} ld\left(\frac{y_1}{R_1} + \frac{y_2^{(m)}}{R_2} \right) \right].$$

$$(4.59b)$$

Equations (4.53)–(4.59) are derived by Smolovich [174]. However, in contrast to [174], the expression given by (4.59) takes into account the complex nature of the refractive index $\hat{n}$. The real part of $\hat{n}$ is close to unity, but the

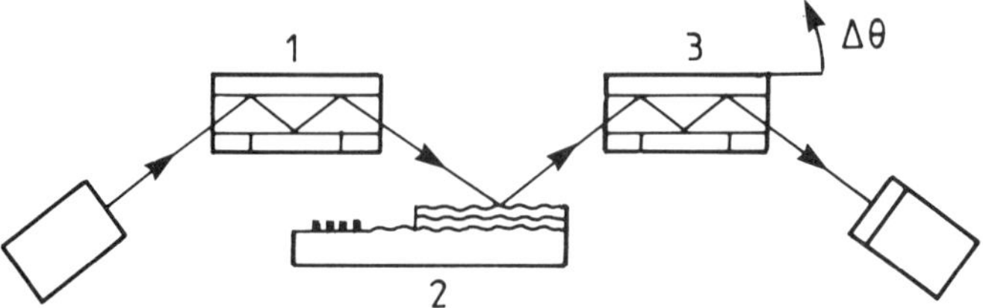

Figure 4.23 Experimental arrangement using a three-crystal X-ray diffractometer: (1) crystal monochromator, (2) modulated multilayer mirror, (3) analysing crystal.

absorption coefficient μ_1 provides a significant contribution to the shape of the reflection curve. The measured diffracted intensity in the mth order is given by

$$I_m\,(x_2,\,y_2) = E_m \times E_m^* = V^2 J_m^2 \left[\frac{2\pi}{\lambda} \tilde{h} \left(\frac{y_1}{R_1} + \frac{y_2^{(m)}}{R_2} \right) \right] G_m \qquad (4.60)$$

$$G_m\,(L) = \frac{1}{L^2} \left\{ \sum_{l=1}^{L} A_{lm} \cos \left[\frac{2\pi}{\lambda} ld \left(\frac{y_1}{R_1} + \frac{y_2^{(m)}}{R_2} \right) \right] \right\}^2$$

$$+ \frac{1}{L^2} \left\{ \sum_{l=1}^{L} A_{lm} \sin \left[\frac{2\pi}{\lambda} ld \left(\frac{y_1}{R_1} + \frac{y_2^{(m)}}{R_2} \right) \right] \right\}^2 \qquad (4.61)$$

$$A_{lm} = \exp \left\{ -\mu_l ld \left(\frac{y_1}{R_1} + \frac{y_2^{(m)}}{R_2} \right) \right\}.$$

In these expressions, the asterisk denotes complex conjugates and G_m describes the selective properties of the multilayer mirror and is effectively the 'excitation coefficient' in the mth diffraction order. The amplitude maxima in zero-order diffraction satisfies the Vul'f–Bragg condition.

The first experimental implementation of the modulation of a soft X-ray beam is reported in [173]. Measurements of the amplitudes and angular distributions of diffracted waves, using a three-crystal X-ray diffractometer (figure 4.23), are reported in [157]. The multilayer structure that was investigated consisted of 400 layers of nickel and carbon with a layer separation of 4.5 nm. It replaced the second crystal after the silicon [111] monochromator. The third crystal, the analyser, was used with an angular resolution of $2''$ to determine the amplitudes fo the components of the spatial spectrum of the diffracted radiation. For a surface acoustic wave of 85 MHz, its wavelength on the surface of the multilayer structure was $\Lambda = 36\,\mu$m. The Bragg angle at the wavelength of 0.154 nm (copper anode) was $0.98°$, so that the angular separation between the zero and first orders was $60''$. According to (4.21), the angular width of the Bragg peak should also be about $60''$ in this case and the diffraction peaks are resolved.

Figure 4.24 (dashed curve) shows the quantities calculated from (4.60). The amplitude of the acoustic wave was measured independently by an optical

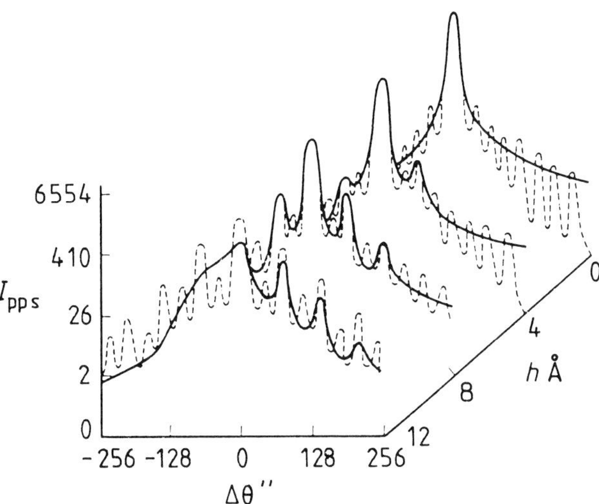

Figure 4.24 Spatial spectrum of the diffracted radiation as a function of the surface amplitude.

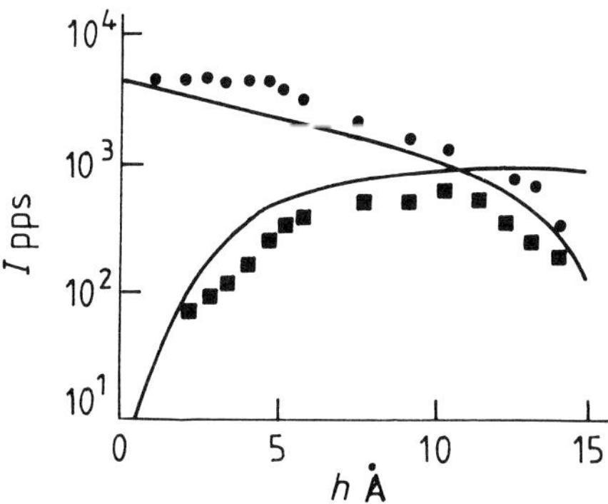

Figure 4.25 Zero- and first-order diffraction intensities as functions of acoustic wave amplitude.

method. It was found that, as the amplitude increased, the energy was transferred from the zero diffraction order to higher orders. Figure 4.25 compares the experimental data with intensity calculations based on (4.60). The discrepancy between calculations and experimental data at the larger amplitudes of the surface waves is due to the heating of the substrate. At the heat released by the interdigitated transducers deforms the surface of the mirror because of the mismatch between the thermal expansion coefficients. The maximum diffracted-wave intensity obtained for an acoustic-wave amplitude of 1 nm was about 20% of the zero-order intensity.

4.4   COMPARISON OF THE PROPERTIES OF TWO- AND THREE-DIMENSIONAL OPTICAL ELEMENTS

The analysis of physical and diffraction principles of X-ray optical elements given in the preceeding sections can be used as a basis for comparing different types of such elements.

Reflecting optical elements based on the phenomenon of total external reflection have obvious advantages such as high efficiency, no chromatic aberration and relatively large aperture. However, the theoretical limit of resolution of grazing-incidence objectives ($\delta_{\min} = \lambda\sqrt{2\delta}$) is practically unattainable for several reasons. Foremost among them is the practical difficulty of producing a smooth surface with a predetermined profile over a large area. Physical considerations show that the estimate $\delta_{\min}$ increases sharply when we take into account the fact that the critical angle of real materials ranges between wide limits. This is due to strong absorption by the surface layer. The experimentally obtained resolution achieved in an image produced by the Kirkpatrick–Baez and Wolter microscopes does not exceed 1–2 μm [37].

A reflecting Bragg system employing spherical multilayer mirrors can be used to develop a more compact objectives at normal angle of incidence. However, their resolution is limited to a few microns because of the same fabrication difficulties and considerable aberrations of the spherical mirrors. Moreover, the efficiency of the Schwarzschild scheme [150] does not exceed 5% (twofold reflection by the multilayer mirrors). The main advantage of the Schwarzschild scheme is its large aperture which means that it is very useful for applications in space.

Fresnel optical elements were developed as a response to the needs of high-resolution X-ray spectroscopy. Their main advantage is that at least in principle, they can produce a resolution of a fraction of a micron. The theoretical limit of resolution for this type of optics is determined by the diffraction conditions for the zone plate and 'diffraction' aberration that arises because of the finite thickness t_{opt} of the zone plate (figure 4.26). If the phase lag of waves diffracted by the entrance and exit surfaces of the optical elements is greater than $\lambda/4$ (the aberration criterion) then strong aberrations will be observed in the image. The condition for the aberration-free image can be derived by analysing, for example, the parabolic zone plate of t_{opt}.

The equation for a zone plate, with the parabolic coordinate as a parameter, can be written in the form

$$r_n^2 = 2F\xi_n^2 + \xi_n^4. \tag{4.62}$$

On the entrance surface (F) of the plate, $\xi_{(1)}^2 = n\lambda/2$ corresponds to the nth Fresnel zone. On the exit surface ($F - t_{\mathrm{opt}}$) an element of the rectangular profile of the zone plate cuts a parabola with parameter $\xi_{(2)n}$. The phase delay corresponds to the difference $\xi_{(1)n}^2 - \xi_{(2)n}^2$, and the Rayleigh criterion gives

$$\xi_{(1)n}^2 - \xi_{(2)n}^2 \leqslant \lambda/4. \tag{4.63}$$

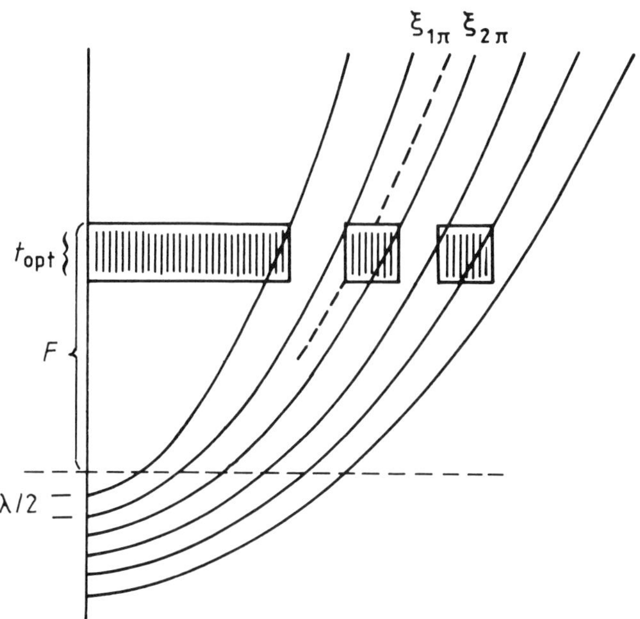

Figure 4.26   Derivation of phase relationships for zone plate with a rectangular profile of finite thickness.

If we suppose that $F \gg t_{opt}$ and neglect quadratic terms in λ in the equation

$$Fn\lambda + n^2\lambda^2/4 = 2\left(F - t_{opt}\right)\xi_{(2)n}^2 + \xi_{(2)n}^4. \qquad (4.64)$$

The condition for aberration-free image transfer signifies that the maximum number of zones in the zone plate cannot exceed

$$n < F/\left(2t_{opt}\right) \qquad (4.65)$$

which corresponds to a minimum zone size in the plate

$$\delta_{min} \geqslant \sqrt{\lambda t_{opt}}. \qquad (4.66)$$

Figure 4.27 shows the maximum theoretically possible resolution of zone plates with rectangular profile of optimum thickness, fabricated from silicon, carbon and gold. Following figure 4.26, the aberrations can be removed by producing a line profile in accordance with the profile of the paraboloids, as indicated by the dashed line. However, this technology is difficult to develop. The currently attained resolution is 50 nm, which is probably close to the limit. All that we have said above will also apply to kinoforms on plate. The kinoform

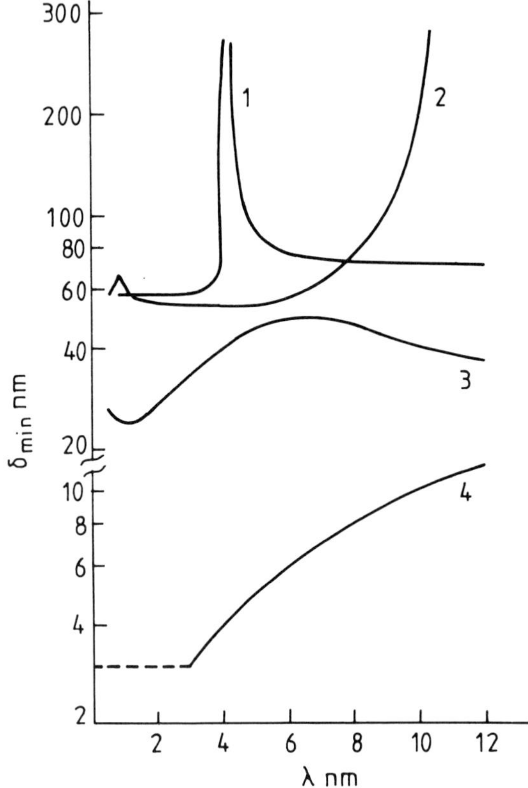

Figure 4.27 Maximum theoretical resolution of zone plates with a rectangular profile for different materials: (1) carbon, (2) silicon, (3) gold, (4) BFL based on Ni/C with a period of 3 nm.

profile for an aberration-free image should also correspond to the profile of three-dimensional Fresnel zones.

The Fourier optics of periodic structures based on the Talbot effect has all the properties of Fresnel optical elements [119].

An important advantage as compared with other methods is the 100% efficiency when coherent radiation is employed. The main disadvantage of this image transfer scheme is the necessity for a periodic multiplication of the image elements. As all other elements of Fresnel optics, the Talbot effect is subject to considerable chromatic aberrations (of the order of λ).

Bragg–Fresnel optical elements combine the high resolving power of Fresnel optics and the high stability of the properties of multilayer mirrors. The use of multilayer coatings in the Wolter scheme [147] results in a resolution of down to $1\,\mu$m. However, the limiting possibilities can probably be realized only in the case of synthesized BFL.

Actually, the resolution of the BFL, even in the case of a rectangular line profile, can reach the diffraction limit of the order of λ. This resolution is attained by profiling the substrate by the amount $\lambda/2$ with subsequent deposition of a multilayer structure. Figure 4.27 shows the calculated limiting resolutions for a carbon phase zone plate and a nickel–carbon biofell as functions of wavelength. For the kinoform BFL, the phase shift depends on the extinction depth t_{ext} and the limiting resolution is given by

$$\delta_{\min} \approx \sqrt{\lambda t_{\mathrm{ext}}}. \tag{4.67}$$

The chromatic aberrations of the BFL are also minimized by three-dimensional diffraction and can be estimated from

$$\delta_{\min} = \frac{\lambda}{2}\sqrt{\frac{F}{2t_{\mathrm{ext}}}}. \tag{4.68}$$

The other properties of the BFL are analogous to those of the Fresnel zone plate.

The optimum depth of the Bragg–Fresnel lens profile. The optimum depth of the lens profile was calculated as follows. Most of the zones on the zone plate have a lateral period Λ greater than the extinction length L_{ext} of the multilayer mirror, which is about $10\,\mu\mathrm{m}$ for the particular Si/W multilayer:

$$\Lambda > L_{\mathrm{ext}} = \frac{t_{\mathrm{ext}}}{\sin\theta_{\mathrm{B}}}. \tag{4.69}$$

The beam is reflected in this case mostly within one period of the grating and ordinary diffraction by the thin grating takes place, producing several diffraction orders within the Bragg peak simultaneously. The optimum profile depth corresponding to maximum first-order diffraction efficiency can be estimated from the formula

$$T_{\mathrm{opt}} = \frac{\lambda^2}{4(\delta_{\mathrm{W}} t_{\mathrm{W}} + \delta_{\mathrm{Si}} t_{\mathrm{Si}})} \tag{4.70}$$

where δ_{W} and δ_{Si} are the optical constants of the multilayer materials, t_{W} and t_{Si} are the corresponding optical paths in the grating materials ($t_i = T_i/\sin\theta_B$), T_i is the thickness of the material.

5

Fabrication techniques

The surge of interest in difffraction elements for X-ray optics that occured in the mid-seventies was due to the advent of new high-resolution techniques in microelectronics technology. It is only since the advent of X-ray and electron lithography, plasmochemical etching and precision techniques for the deposition of thin films that it has become possible to fabricate diffracting structures with dimensions down to a few hundred ångströms. The new X-ray diffraction optics has in turn served as an excellent testing ground for the development of the latest microelectronics technology with nanometre resolution. When compared with traditional electronics, diffraction optics must meet a number of stringent technological conditions in relation to the precision with which the topological elements are reproduced. The phase-sensitive character of X-ray optical devices determines the relative precision of the structure elements. For example, if the required spatial resolution of a zone plate is $0.3\,\mu\mathrm{m}$, the ring diameter of the last zone must be accurate to within $0.1\,\mu\mathrm{m}$, i.e., 0.02% for an aperture of $500\,\mu\mathrm{m}$. The wavelength of the X-rays employed is 0.1–$10\,\mathrm{nm}$ which means that the necessary technology must have comparable spatial resolution.

The fabrication of X-ray diffraction elements often relies on a controlled resist mask profile for the special technological procedure employed, e.g., lift-off lithography. In general, the topology of a diffraction element is a complicated structure, described by second- or third-degree curves, which is very unusual in traditional microelectronics. There is therefore a requirement for a flexible method of producing the required topology with the possibility of analytic description and correction of any attendant distortions. Only a dedicated electron-beam lithography system is capable of meeting such stringent conditions.

All these factors had to be taken into account in the design of the dedicated complex of technologies and electron-beam lithographic equipment for the fabrication of X-ray diffraction optics, optoelectronics and acoustoelectronics described in [175]. The lithographic complex was developed at the Institute of Microelectronics Technology of the Russian Academy of Sciences between 1980 and 1990 and was used to fabricate the X-ray optical elements described in this book.

5.1 FABRICATION TECHNIQUES FOR DIFFRACTION ELEMENTS

We shall now examine alternative electron-beam lithographic techniques for fabricating X-ray diffraction optics with the required topologies.

Holographic laser lithography. The most highly developed technique for producing diffraction gratings and zone-plate optics is the holographic method [176, 177]. This relies essentially on the utilization of the interference pattern produced by two coherent laser beams in the visible or ultraviolet ranges. However, very considerable aberrations arise when this *hologram* is read by a beam of wavelength that is shorter by two orders of magnitude than the wavelength used to write the hologram. The phenomenon was originally considered by Gabor who was the first to propose the holographic method [178]. In the simplest situation, e.g., in the case of the hologram of a point, the problem can be solved by compensating aberrations with aspherical waves during the writing process. As described in [177], the fabrication of a zone plate by the holographic method consists of the following stages:

(1) Design of the optical arrangement for writing the hologram, which ensures that aberrations are corrected when optical radiation is replaced with X-rays.

(2) Exposure of the photoresist layer deposited on a glass substrate and coated with a polymer and gold.

(3) Development of the resist, ion-beam etching of gold through the resist mask and removal of the glass substrate.

This produces a zone plate consisting of a gold mask on a thin polymer membrane that is transparent to X-ray radiation.

Zone plates in which the size of the last zone is down to $0.055\,\mu$m and the number of zones is of the order of a few hundred can be fabricated using ultraviolet radiation with a wavelength of 257 nm (second harmonic of the argon laser with $\lambda = 514.5$ nm). By copying this zone plate in soft X-rays by the interference method, it is then possible to reduce the size of the last zone to $0.018\,\mu$m. However, although this is possible in principle, the technique encounters considerable experimental difficulties because, so far, no-one has succeeded in fabricating zone widths of less than $0.055\,\mu$m by the holographic method.

Zone plates produced by the holographic method at Göttingen University have demonstrated the very considerable possibilities of X-ray microscopy in experiments performed on the X-ray sources in Berlin (BESSY) and Orsay (ACO) [179,180]. The main difficulty in producing microzone plates by the holographic method is that the thickness of the resist on the last zones must amount to only a few dozen nanometres (cosinusoidal distribution of the resist exposure). This means that the last line cannot be produced in the metal with sufficient absorbing-layer thickness (aspect ratio). Zone plates made by the holographic method have low contrast and images produced with them require careful *a posteriori* processing. Other disadvantages of the holographic method include the lack of topological flexibility, e.g., in producing ellipsoidal zones. The

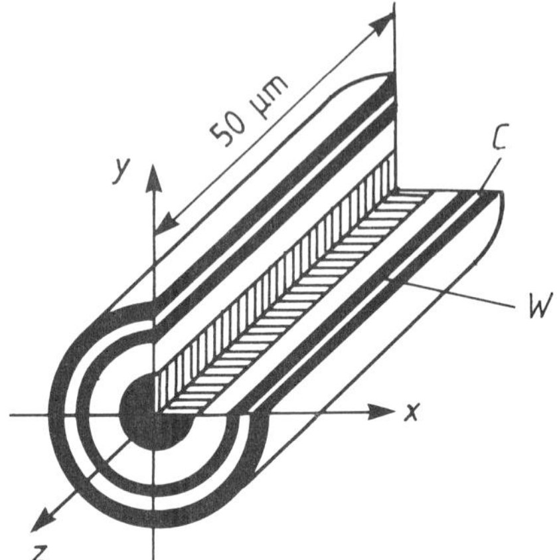

Figure 5.1   The layered zone plate consisting of Si and W [182].

holographic method probably finds its optimum application in the fabrication of condenser zone plates used for the preliminary monochromatization of radiation and the concentration of the flux on the object being observed [177]. Condenser elements must have large diameter and relatively long focal length. Consequently, the number of zones on such plates must reach a few thousand. Obviously, this cannot be achieved by other lithographic techniques. The scheme developed at Göttingen University for producing condenser zone plates can be used to fabricate zone plates with 15 000 zones and minimum width of 0.15 μm. The aberration-limited minimum spot size produced by such zone plates is 3–5 μm.

Layer-by-layer deposition. 'Sliced' zone plates [181] have a very promising future. In this method, the zones are produced by the successive deposition of materials with low and high absorption coefficient on a rotating wire. The resulting multilayer 'roll' is then sliced into individual zone plates. One of the basic problems in this technique is the production of a wire whose diameter can be kept constant to within 10 nm. Ion sputtering on to a rotating target has been suggested for the deposition process itself. This type of technology produces layers whose thickness can be maintained with relative precision of 10^{-4}. The main difficulty lies in slicing the 'roll' into the individual zone plates. The thickness of less than 1 μm that is required for the nanometre wavelength range is exceedingly difficult to achieve. Recent publications report tests on zone plates fabricated by this technology, using hard X-rays (8 keV) [182]. The zone plates consist of 20 pairs of silicon and tungsten layers and their overall thickness is 50 μm (figure 5.1). The considerable length of the zone plate in this case means that the conditions that have to be satisfied by the roll slicing procedure are less stringent, but the spatial resolution is still limited to something of the order of 1 μm. A spot size of 5 μm was achieved in the experiment published in [182].

Ion-beam lithography. Increasing numbers of publications on ion-beam lithography with a focused ion beam have appeared since 1980 [183–187]. This has been due to advances in, and industrial applications of, liquid-metal ion sources, precision focusing systems and ion-beam transport control. In their general design, lithographic systems incorporating ion sources are are similar to electron-beam lithographic systems. Their main advantages are as follows:

(a) No proximity effects during the interaction of ions with the resist and the substrate.

(b) Localized regions can be doped by the ion beam method.

(c) The surface can be directly etched with a submicron ion beam.

The development of ion-beam lithography (IBL) was pioneered at the Centre for Nanolithography at Osaka in Japan where a whole range of ion lithographs for laboratory investigation has been fabricated. The British firm VG-Ionics undertook, almost simultaneously, the production of the first commercial system for ion-beam lithography (the IBL-100) which uses a 100 keV probe and a liquid-metal gallium ion source. At present, ion-beam lithography is still in the laboratory stage, but the number of publications in this area is rapidly increasing. Research is also continuing in France [186] and at Cornell University in the United States [184].

Tunnelling lithography. A new lithographic technique based on the use of the scanning tunnelling microscope is beginning to develop [188]. The resolution of this process can reach 1.0 nm and initial results suggest that this is a realistic figure. Only a few papers have appeared in this field, but the method may well turn out to be very useful in X-ray optics technology.

The above lithographic methods are suitable for the fabrication of diffraction optics elements and, like the electron-beam lithography discussed below, can be used to produce single examples of unique structures. Before practical applications become possible, searches will continue for suitable methods for short-run fabrication of elements with reproducible parameters. Among promising lithographic methods that are being developed for industrial microelectronics, the highest resolution has been achieved in those employing ultraviolet, vacuum ultraviolet [189] and X-ray radiation.

Ultraviolet and vacuum ultraviolet lithography can be used for short-run fabrication of condenser zone optics and optical diffraction systems for ultraviolet projection lithography which requires a resolution of 0.25–0.3 μm. The principle of X-ray diffraction optics can readily be extended to the ultraviolet and vacuum ultraviolet regions which opens a new stage in the development of lithographic systems. X-ray lithography is regarded as one of the most promising technological techniques for the next century and is likely to replace optical lithography [190–197]. The development of X-ray lithography is actively continuing in many countries across the globe.

5.2 ELECTRON-BEAM LITHOGRAPHY IN THE FABRICATION OF X-RAY OPTICAL ELEMENTS

5.2.1 Limiting resolution of electron-beam systems

Electron-beam lithography (EBL) is at present the main method of generating the images of topologies with submicron element dimensions. There are several monographs on this subject [191–193], so that we shall concentrate our attention on the role of EBL in the fabrication of the elements for X-ray optics.

The de Broglie wavelength of electrons of energy W is given by

$$\lambda_e = (150/W)^{1/2} \tag{5.1}$$

where W is in keV and the wavelength is in nanometres, so that for energies in the range 10–100 keV the wavelength lies in the range 0.12–0.4 nm. Diffraction phenomena at such wavelengths are relatively unimportant and have no effect on spatial resolution [198]. The main parameters that affect spatial resolution in electron lithography are the probe size and the accelerating voltage, and the two are related.

The minimum probe size depends on the aberrations of the electron-optical system, the probe current and the accelerating voltage [199]:

$$d_p = C_s \alpha_{\mathrm{opt}}^3 \tag{5.2}$$

where α_{opt} is the optimum electron collection angle within the aperture of the electron lens and C_s is the spherical aberration coefficient of the electron lens. In the case of thermionic emission from a lanthanum hexaboride cathode, the optimum angle is given by

$$\alpha_{\mathrm{opt}} = 0.93 i_0^{1/8} B_e^{-1/8} C_s^{-1/4} \tag{5.3}$$

where i_0 is the electron probe current and B_e is the the cathode brightness.

The brightness of the electron gun in $\mathrm{A\,cm^{-2}\,sr^{-1}}$ depends on the type of cathode and the accelerating voltage:

$$B_e = 2.5 \times 10^6 J_e W_e / \pi k_B T_e \tag{5.4}$$

where J_e is the emission current density, W_e is the electron energy in eV, T_e is the effective temperature of the cathode and k_B is the Boltzmann constant. In the case of tungsten and La_2B_6 thermionic cathodes, the values of B_e at 30 keV are 4×10^4 and $2.5 \times 10^2 \mathrm{A\,cm^{-2}\,sr^{-1}}$, respectively.

It follows from (5.2)–(5.4) that

$$d_p = (J_e / W_e)^{3/8} \,. \tag{5.5}$$

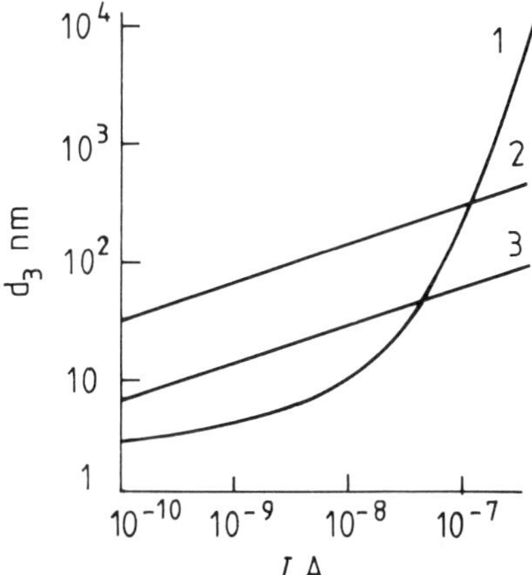

Figure 5.2   Theoretical dependence of the electron-probe diameter on the beam current [199]: (1) field-emission gun, (2) thermionic gun, (3) La_2B_6 gun.

This shows that the probe size is proportional to the current and inversely proportional to the electron energy, both raised to the power 3/8.

It follows that, to achieve maximum resolution, the current must be as low as possible and the accelerating voltage as high as possible. However, in practice, the beam current cannot be reduced too much because the dose necessary for the exposure of the resist lies in the range 0.01–1 J cm^{-3} and 10^{-1}–10^2 J cm^{-3} for positive and negative resists, respectively. The beam current for the optimum exposure time must therefore lie in the range 10^{-9}–10^{-8} A. It is important to note that the formula given by (5.5) was derived for a thermionic emission cathode. In the case of the field emission gun, the probe size is a different function of the beam current:

$$d_p = i_0^{3/2}. \tag{5.6}$$

Figure 5.2 shows the probe diameter as a function of the beam current for three types of electron gun, namely, field emission, thermionic emission and La_2B_6 at electron energy of 30 keV [199]. It is clear from this figure that the probe size in the case of the field emission gun is of the order of 10 nm up to beam currents of 10 nA. From this point of view, the field emission cathode is the best solution for high-resolution EBL. The best resolving power has indeed been achieved by using field emission cathodes [200–203]. Examples of systems of this kind are the modified scanning transmission electron microscopes HB-5 and HB-501 manufactured by the Vacuum Generator Company. The systems are being successfully used to fabricate unique experimental structures and zone plates with nanometre zone widths [204–207].

EBL holds out the possibility of microelectronics based on completely new fabrication principles. High-voltage EBL with a beam of circular cross section and probe energy of 50–100 keV can be used to fabricate structures with element dimensions down to 8 nm [208]. This has now been designated

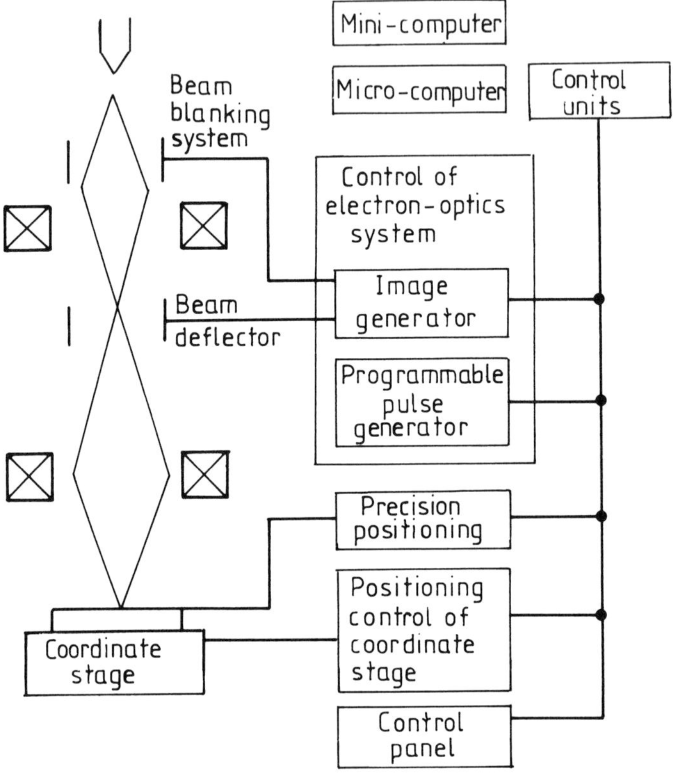

Figure 5.3   Block diagram of an electron-beam lithographic system.

as 'nanolithography'. Apart from the fabrication of diffracting structures, nanolithography has also found applications in computer technology requiring elements with dimensions down to 10–100 nm. Application areas include quantum-mechanical devices, molecular electronics, ballistic electronics and so on.

5.2.2   Implementation of electron-beam lithography

The electron-beam lithographic system is a complicated apparatus that consists of the following main components (figure 5.3):

(1) Electron-optical system, source of electrons and beam blanking system.

(2) Coordinate stage with precise positioning and a laser interferometer used to monitor the position of the stage.

(3) Minicomputer that controls the operation of all systems.

Other essential components of the electron-beam lithographic system are a vacuum system, stabilized power supplies and various monitoring devices.

Depending on the principles used to draw the required pattern, EBL systems can be classified as scanning systems, vector scanning systems, systems with programmed beam profile, projection and multibeam systems [191–193, 209].

EBL systems with an electron beam of circular cross section and based on the vector scanning principle have very low fabrication efficiency. However, the structure dimensions are accurately reproduced by these systems as compared with the more productive vector systems employing a programmed beam profile. High-intensity electron beams are successfully used in vector scanning systems to fabricate structures by resist-free technology. In [210], zone plates with line widths down to 15 nm were produced by direct carbonization of a thin oil film by a high-intensity electron beam. The high-resolution lithographic process used in [211] was based on the decomposition of metallorganic compounds by an electron beam. Electron-beam-induced chemical reactions have been used to produce copper diffraction gratings in polyvinyl alcohol with submicron resolution [212].

We shall now consider the EBL system used in [67, 81, 153, 158, 162, 164] to fabricate phase zone plates and Bragg–Fresnel elements. The ZRM-12 system manufactured by Carl Zeiss Jena was used as the basic system. (Now Jenoptik.)

Figure 5.4 shows a schematic diagram of the electron-optical part of the system. The beam of electrons produced by the electron gun is focused on a substrate coated with a resist. Computer-controlled deflection and blanking systems are used to expose particular regions on the resist and thus produce the required topology. Step motors in the stage-displacement system ensure precise transport of the plate in the horizontal plane in two directions. A laser interferometer is used to monitor the position of the stage. Typical measurement accuracy is ± 0.1 to $\pm 0.04\,\mu$m (ZRM-12, 20).

The size of the working field (without stage displacement) is limited by distortion in the electron-optical system. It is chosen in the light of the required structure fabrication precision, in accordance with the properties of the electron optics system and the beam step size which is computer controlled via analogue-to-digital converters (DAS). The beam step size achieved with a 14-bit DAC and an exposed field of 1 mm $\times$ 1 mm is $0.0625\,\mu$m.

As already noted, the size of the electron probe is one of the basic parameters of the system. The limiting possible probe diameters are given by (5.2) and (5.5).

Figure 5.6 shows the measured size of the probe in the ZRM-12 system for different values of the beam current. The probe size was determined by measuring the secondary-electron emission profile for the image of a perfect step (figure 5.5). The test object was the cleavage plane of a silicon crystal. The probe size d_p is related to β_k (figure 5.5) by:

$$d_p = 1.6\beta_k/M$$

where M is the magnification.

Figure 5.6 shows the probe size as a function of the beam current for the ideal and actual beam systems (ZRM-12 with probe energy of 30 keV). Reliable

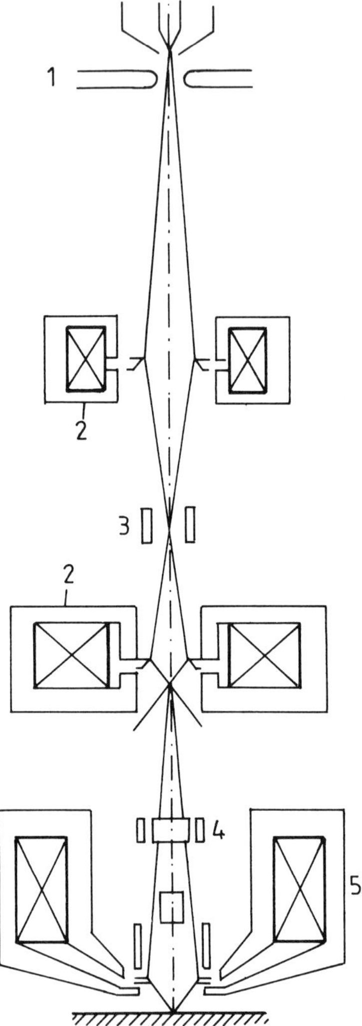

Figure 5.4 Schematic diagram of the electron-optical part of an EBL system: (1) electron gun incorporating the La_2B_6 cathode, (2) condenser lenses, (3) blanking and deflecting plates, (4) deflecting system, (5) objective lens.

experimental values of d_p cannot be obtained for probe currents below 10^{-10} A because of signal noise. The actual probe diameter in an electron-beam system incorporating a thermionic-emission cathode is 50–100 nm (25–500 nm in the case of the La_2B_6 cathode) for an accelerating voltage of 30 kV. The width of structures produced in a resist is greater by a factor of five to 10. A strong interaction is observed between closely spaced structures, giving rise to the

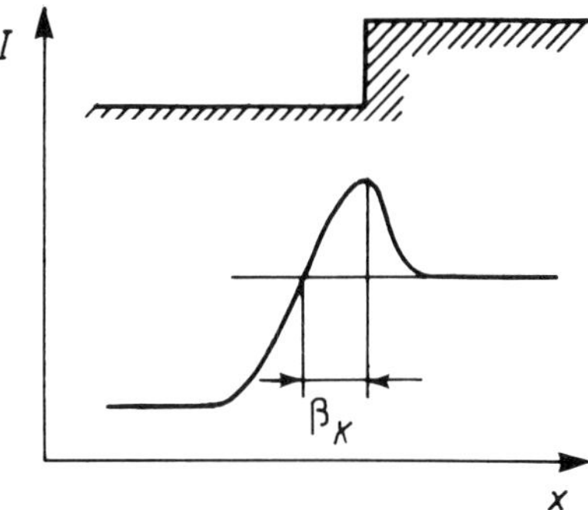

Figure 5.5 Signal from the secondary-electron detector obtained for a test object in the form of a step.

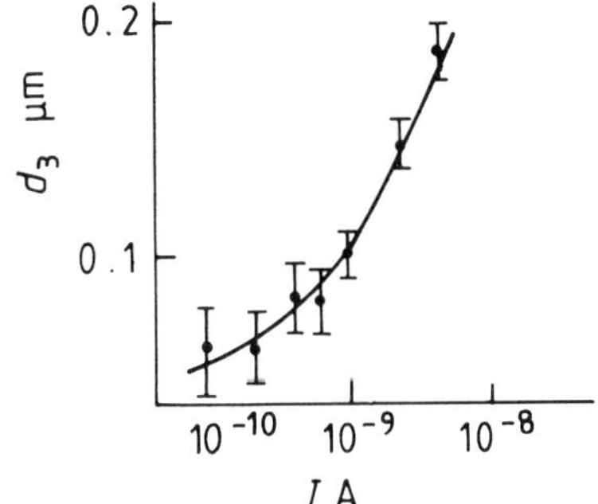

Figure 5.6 Diameter of the electron probe as a function of the beam current in the ZRM-12 system. Solid line is calculated.

magnification and distortion of their dimensions.

A small-diameter electron probe is necessary, but not sufficient for the production of submicron structure dimensions. Beginning with widths of 0.5–1 μm, the image is broadened mostly by secondary processes associated with the dissipation of the incident-electron energy in the resist and in the substrate. They are referred to as proximity effects and describe in the general case the three-dimensional distribution of primary electron energy absorbed in the resist.

5.2.3 Distortion of structure dimensions in EBL

It has been demonstrated experimentally that this effect has a range of 3–5 μm for 30 keV electrons in polymer resists. The phenomenon can be described in terms of two basic processes, namely, back scattering of primary electrons by the substrate and the development effect (processing of the resist). The topic has attracted a large number of papers and monographs [213–225]. There is experimental evidence for the simultaneous presence of several phenomena, and we shall therefore consider the the integrated effect on structure dimensions in a system with a Gaussian energy distribution in the beam and a circular focal spot [223]. If we take the probe diameter d_p as the characteristic beam size, the differential electron-flux distribution on the surface of the resist is

$$dP(r) = \frac{P_0}{\pi d_p^2} \exp\left(-\frac{r^2}{d_p^2}\right) r \, dr \tag{5.7}$$

where r is the distance from the beam axis and P_0 is the maximum intensity of the incident beam. The incident electron beam is scattered as a result of interaction with the substrate material and produces a number of secondary photons and electrons, including true secondary electrons with energies up to 50 eV, reflected electrons with energies close to the primary-beam energy, secondary electrons with energies up to 500 eV, secondary X-rays, cathodoluminescence and so on. These processes have different efficiencies and measurements have shown that the main contribution to the resist exposure is provided by electrons with energies up to 3 keV [224]. Secondary X-rays can be neglected. Bearing in mind the physical picture of the process, the differential scattering function has the following form at each point:

$$\Phi(r) = \frac{dP(r)}{\pi Q_0} \sum_{i=1}^{\infty} \frac{\eta_i}{\alpha_i^2} \exp\left(-r^2/\alpha_i^2\right) \tag{5.8}$$

where Q_0 is a normalizing factor, η_i is the relative contribution of each component of the process to the resist exposure and α_i is the geometrical parameter of the scattering function corresponding to the ith process.

Assuming that the main contribution is due to forward and back scattering of primary-beam electrons, we find that the scattering function consists of two terms, namely,

$$\Phi(r) = \frac{dP(r)}{\pi(1 + Q)} \left[\alpha_1^{-2} \exp(-r^2/\alpha_1^2) + \eta_2 \alpha_2^{-2} \exp -r^2/\alpha_2^2\right] \tag{5.9}$$

where $\eta_1 = 1$, $Q = 1 + \eta_2$ and η_2 is the back-scattering coefficient.

The convolution of the original beam distribution and the scattering function then gives

$$D(r) = \frac{P_0 \tau_0}{\pi(1 + Q)} \left[\frac{1}{\alpha_1^2 + d_p^2} \exp\left(-\frac{r^2}{\alpha_2^2 + d_p^2}\right) \right.$$
$$\left. + \eta \frac{1}{\alpha_2^2 + d_p^2} \exp\left(-\frac{r^2}{\alpha_1^2 + d_p^2}\right) \right] \tag{5.10}$$

where τ_0 is the structure exposure time. It was assumed in the derivation of this formula that the energy density absorbed by the resist was proportional to the flux density of electrons in the primary beam. X-ray optical elements usually have a linear structure or a structure described by second-degree curves with relatively large radius of curvature, so that knowledge of the linear scattering function is sufficient in most cases. To transform from the point proximity function to the corresponding linear case we have to take into account the discrete step size of the exposure system and sum over all steps. The linear distribution function for the absorbed energy density [223] then becomes

$$D(x) = \frac{P_0 \tau_0 \sum \alpha_1'}{\pi(1 + Q)(\alpha_1^2 + d_p^2)} \left[\exp\left(-\frac{x^2}{\alpha_1^2 + d_p^2}\right) + C \exp\left(-\frac{x^2}{\alpha_2^2 + d_p^2}\right)\right]$$

(5.11)

where

$$C = \eta_2 \frac{(\alpha_1^2 + d_p^2) \sum \alpha_2'}{(\alpha_2^2 + d_p^2) \sum \alpha_1'}$$

$$\sum \alpha_1' = \sum_{n=1}^{\infty} \exp\left(-\frac{n_1^2 \Delta y^2}{\alpha_1^2 + d_p^2}\right) \qquad \sum \alpha_2' = \sum_{n=1}^{\infty} \exp\left(-\frac{n^2 \Delta y^2}{\alpha_2^2 + d_p^2}\right)$$

and Δy is the beam step size along the y axis.

The distribution of absorbed energy density over a structure that is infinite in the y direction therefore becomes

$$D(x) = \frac{P_0 \sum \alpha_1'}{\pi(1 + Q)(\alpha_1^2 + d_p^2)}$$
$$\times \sum_{i=1}^{\infty} \tau(x_i) \left\{ \exp\left[-\frac{(x - x_1)^2}{\alpha_1^2 + d_p^2}\right] + C \exp\left[-\frac{(x - x_1)^2}{\alpha_1^2 + d_p^2}\right] \right\} \quad (5.12)$$

where P_0 is the peak electron flux density and $\tau(x_i)$ is the exposure time at the ith point. The sum is evaluated over all the x_i with the step size Δx. The general form of (5.12) is

$$D(x) = P_0 \sum_{i=1}^{\infty} \tau(x_i) \Phi_{\text{lin}}(x - x_i)$$

(5.13)

where

$$\phi_{\text{lin}}(x) = \frac{\sum \alpha_1'}{\pi(1 + Q)(\alpha_1^2 + d_p^2)} \left[\exp\left(-\frac{x^2}{\alpha_1^2 + d_p^2}\right) + C \exp\left(-\frac{x^2}{\alpha_2^2 + d_p^2}\right)\right]$$

is the linear image spread function corrected for the discrete exposure step.

5.2.4 Algorithm for the correction of proximity effects

The distortion of the structure in the resist that occurs after exposure and processing is determined by the following main factors: (1) effects associated with primary-beam scattering in the resist and substrate, (2) the influence of the development process and (3) the discrete character of the motion of the beam [225]. It has been demonstrated experimentally that the error in the exposure dose for an element with linear dimensions of $0.3\,\mu$m should not exceed 20%.

The exposure time $\tau(x_i)$ necessary to produce a given absorbed-dose distribution can be determined from (5.13) by a relatively simple iteration procedure. The zero-order approximation to the exposure time for the structure is naturally taken to be

$$\tau^{(o)}(x) = D(x)/P_0 \tag{5.14}$$

i.e., without the proximity effects. The successive approximate values $\tau^{(k)}$ can be found from the iteration formula

$$\tau^{(k+1)}(x) = \frac{D(x)\tau^{(k)}(x)}{\sum\limits_{i=1}^{\infty}\tau^{(k)}(x)\phi_{\mathrm{lin}}(x - x_i)}. \tag{5.15}$$

The next step towards higher precision in the fabrication of the structure profile is to take into account the development process, i.e, to determine the exposure time $\tau > 0$ for which the resist assumes a given form $Z(x)$ after development time $\tau_{\mathrm{dev}}(x)$. A model of local isotropic development of polymer resists is discussed in detail in [226, 227]. The iteration algorithm discussed in [228, 229] can be used to allow for the development process in terms of a correction factor. Moreover, this iteration scheme automatically eliminates negative values and requires very little computer memory.

5.3 DISSIPATION OF THE PRIMARY-ELECTRON ENERGY

5.3.1 Effect of the physical properties of the substrate

The parameters of the function describing the dissipation of the primary-electron energy are determined by the physical properties of the substrate and resist materials. When the physical principles underlying the fabrication of X-ray optics elements are examined, it is important to investigate how the exposure of process is affected by secondary emission, the energy spectrum of back-scattered electrons, their angular distribution and also the charge-transfer processes. Preliminary estimates of the properties of the material can be based on the Grun formula [230] which gives the range of electrons as a function of the density of the medium and the primary-beam energy:

$$r_{\mathrm{e}} = 0.046W^{1.75}/\rho \tag{5.16}$$

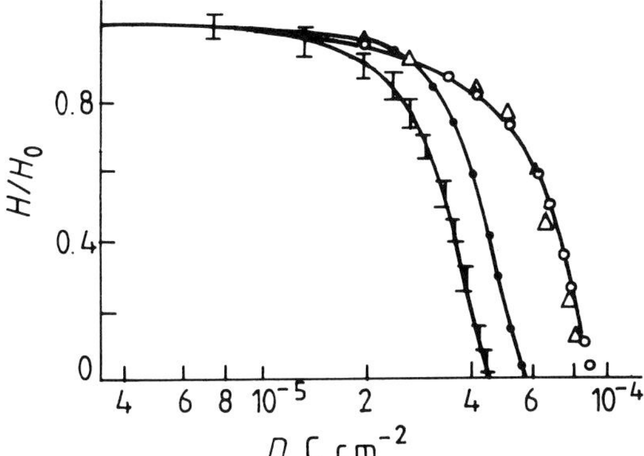

Figure 5.7   Dose characteristics of an electron resist on substrates of silicon (I), lithium niobate ($\bullet$), polyimide on silicon ($\triangle$) and polyimide ($\bigcirc$) [224].

where r_e is the electron range in microns, ρ is the density in $g\,cm^{-3}$ and W is the probe energy in keV. It is clear from this formula that the electron range in light materials is quite large, so that electrons can traverse long distances without much scattering.

This effect is flash in the technology used to fabricate transparent zone plates on thin membranes because (5.16) shows that the range of 30 keV electrons in polyimide is approximately $13\,\mu m$. When the membrane thickness is $1\,\mu m$, approximately 99% of the primary-beam electrons succeed in passing through it.   The contribution to exposure due to back-scattered electrons is so small that it can be neglected. True secondary electrons leaving the membrane with energies of a few eV have ranges of $10\,nm$ and do not affect the exposure process [224]. Consequently, the exposure of the resist is entirely due to primary-beam electrons when the polyimide membrane is used as the substrate.

The same effect can be achieved by using polyimide between the silicon substrate and the resist (two-layer resist).   Figure 5.7 shows the dose characteristics of the polymethyl methacrylate resist (PMMA), $0.3\,\mu m$ thick, deposited on a self-supporting polyimide membrane, a polyimide membrane on a silicon substrate, a pure silicon substrate and a $LiNbO_3$ substrate with a titanium film [224]. It is clear from figure 5.7 that the intermediate polyimide layer with a thickness of $1.0\,\mu m$ suppresses the influence of the substrate almost completely. The dose characteristic can be used to determine the contribution of back-scattered electrons to the energy absorbed by the resist. The sensitivity of the resist determined from curves 1, 2 and 3 in figure 5.7 is not the same, and it is clear that this difference is due to electrons leaving the substrate.

The contribution of this process to the overall irradiation dose can be estimated from the formula

$$\Delta D = (D_1 - D_0)/D_0 \qquad (5.17)$$

where D_0, D_1 are the doses necessary for the development of the resist down to the membrane and to the substrate, respectively.

For the silicon substrate, the contribution of back-scattered electrons was $\Delta D = 0.51$. This means that, for given exposure time and beam current, the topology elements on the silicon substrate receive an exposure that is greater by a factor of two than on the polyimide membrane. The dose characteristics shown in figure 5.7 can be used to estimate the energy of electrons that provide the maximum contribution to the additional exposure dose. Actually, when the polyimide layer thickness is $0.4\,\mu\text{m}$, and the layer is deposited on the silicon substrate, the results of exposure are no different from those for a self-supporting membrane. High-energy electrons lose only a small fraction of their energy as they cross the polyimide. The specific energy loss for a beam of 20 keV electrons is $2\,\text{eV}\,\text{nm}^{-1}$. The total loss in a layer of $0.4\,\mu\text{m}$ is 800 eV. Only electrons with energies up to 5 keV experience appreciable attenuation in the film. Hence it may be concluded that it is precisely these electrons that provide the main contribution to the proximity effect.

However, we do not always have the freedom to choose the substrate (examples include the fabrication of Bragg–Fresnel elements and X-ray modulators). The fabrication of acoustic X-ray modulators by electron-beam lithography is particularly difficult [173–174]. A surface acoustic wave (SAW) can be excited on a substrate by a special periodic metallized structure (interdigital transducer) formed on the surface. High-frequency SAW devices employ electrodes of width less than $1\,\mu\text{m}$, separated by a similar distance. These devices are usually deposited on piezoelectric materials such as quartz, lithium niobate and so on. Since the substrate is a dielectric, it accumulates negative charge. This effect can be partially suppressed by depositing a metal film on the surface of the crystal. However, piezoelectric materials also exhibit spontaneous polarization. In the case of $LiNbO_3$, the penetration depth of 30 keV electrons is about $2\,\mu\text{m}$. These electrons create polarized regions that affect the scattering of primary-beam electrons even through a metal film [232].

This phenomenon may increase or suppress the proximity effect, depending on the initial direction of the spontaneous polarization vector in the crystal. For example, it is shown [232] that lithography on the YZ cut of an $LiNbO_3$ crystal in which the spontaneous polarization vector P_s points into the crystal produces a smaller back-scattering effect (as compared with the silicon substrate). A study reported in [233] has shown that the deposition of the metal film can be avoided by performing the lithography on a $127°$ YZ cut in $LiNbO_3$ crystals. The pyroelectric properties of the $LiNbO_3$ substrate can be used to produce a temperature gradient in the interior of the crystal and thus suppress almost completely the proximity effect by means of the internal electric field. Actually, the primary-electron probe heats up the substrate at the point of exposure of the structure by the amount

$$\Delta T = P/4.27 d_p q_T \tau \qquad (5.18)$$

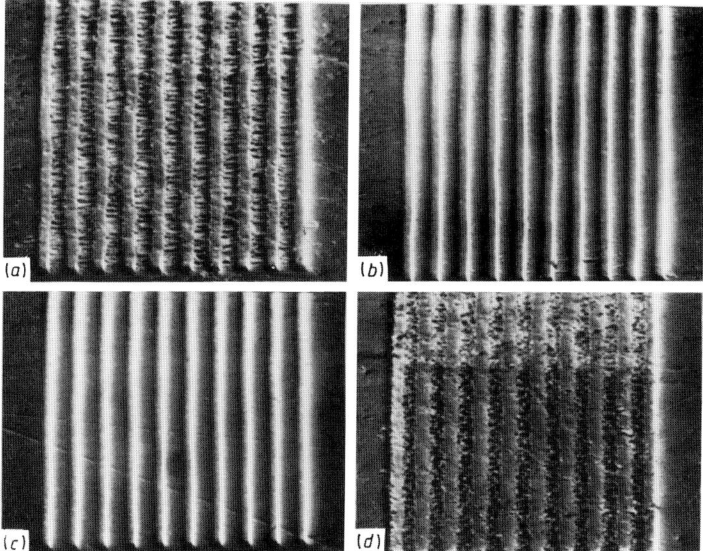

Figure 5.8 Micrograph of a test structure with a period of $2\,\mu$m, produced by direct exposure to the electron beam of the surface of a lithium niobate crystal covered by a negative electron resist [233]. The values of ΔT were as follows (in K): (a) 0, (b) -5.4, (c) -10.7, (d) 14.2.

where ΔT is the rise in the absolute temperature of the substrate due to the electron probe, P is the electron probe power in W, d_p is the electron probe diameter in cm, q_T is the thermal conductivity ($0.46\,\mathrm{J\,m^{-1}s^{-1}K^{-1}}$) [234] and τ is the exposure time in seconds.

The temperature rise ΔT in the substrate, associated with to the pyroelectric effect, produces a negative surface charge with density given by

$$\Delta Q_i^s = q_i \Delta T \tag{5.19}$$

where q_i is the pyroelectric coefficient (for $LiNbO_3$ at 300 K, $q_1 = 0, q_2 = 0, q_3 = -4.8410\,\mathrm{C\,m^2\,K^{-2}}$) [234].

Exposure to a dose of $0.02\,\mathrm{C\,m^{-2}}$ produces a 15 °C rise in the temperature of the substrate. According to (5.19), the charge density for the 127° orientation should be $-5 \times 10^{-6}\,\mathrm{C\,m^{-2}}$ [233]. The total negative surface charge density introduced by the electron beam and induced by spontaneous polarization leads to a change in secondary-electron emission from the substrate and to a defocusing of the image of the lithographed structure (figure 5.8(a)). A compensating positive pyroelectric charge $Q^s = 2.1 \times 10^{-4}\,\mathrm{C\,m^{-2}}$ can be produced on the surface of the crystal by coooling it by 5.4 K with a Peltier microjunction. This completely compensates the pyroelectric charge produced by the beam and, to some extent, the charge introduced by the beam itself (figure 5.8(b)). When the temperature difference is raised to 10.7 K, the

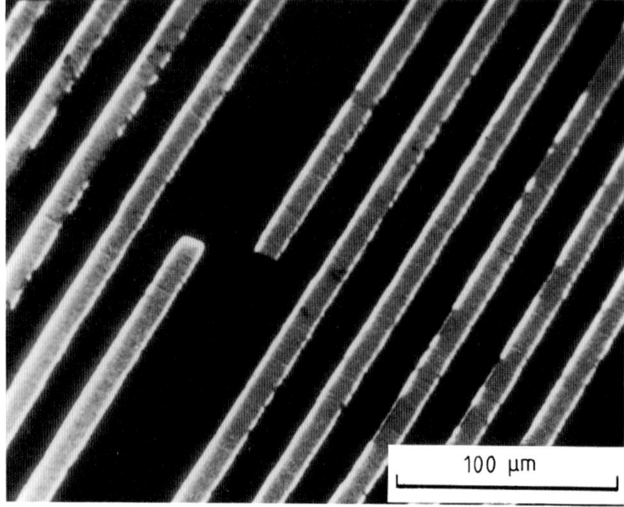

Figure 5.9 Fragment of the structure of the interdigital transducer working at 580 MHz on the surface of LiNbO$_3$ [235].

proximity effect is almost completely suppressed ($Q^s = 4.1 \times 10^{-4}\,\mathrm{C\,m^{-2}}$) and a stable positive charge is produced on the surface of the crystal (figure 5.8(c)). For comparison, figure 5.8(d) shows the same topology exposed whilst the underside of the crystal was heated. The appearance of a high negative charge produces the complete spreading of exposed structure. Figure 5.9 shows a fragment of the interdigital transducer used to excite the surface acoustic wave at 580 MHz.

5.3.2 Measurement of the energy-dissipation parameters in an electron resist

To calculate the actual distribution of energy absorbed by a resist, we must know the numerical values of d_p, α_1 and α_2. The diameter of the electron probe d_p can be measured independently by the procedure described above. A large number of different models has been proposed for the determination of the parameters α_1 and α_2, but most of them rely on the availability of a relatively arbitrary initial data. The method described in [233] is free from these disadvantages and yields these parameters in an explicit form without any predetermined conditions. It is based on the following principles. A rectangular area on the substrate covered with the resist is given a linearly increasing exposure dose ranging from zero to D_0, where the latter corresponds to the development of the resist down to the substrate (figure 5.10(a), (b)). A strip of minimum thickness is exposed over the rectangle with a constant of dose. The combination of absorbed energy densities due to the rectangle and the strip broadens its thickness at the higher doses on the left of the rectangle

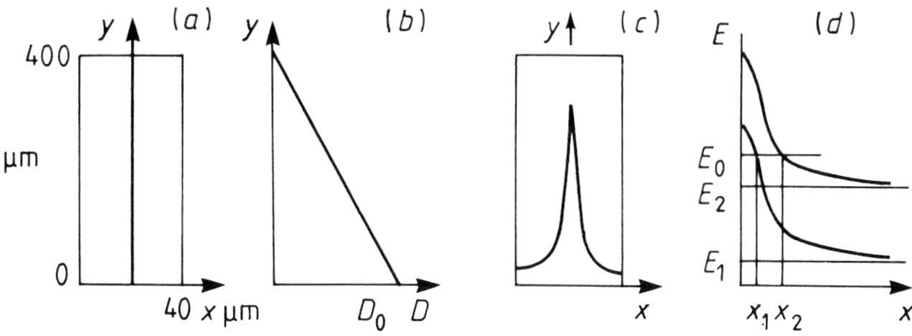

Figure 5.10 Interaction between an electron probe and a substrate: (*a*) test structure, (*b*) distribution of dose over the test structure, (*c*) test pattern after development, (*d*) half-width of the developed region x_2.

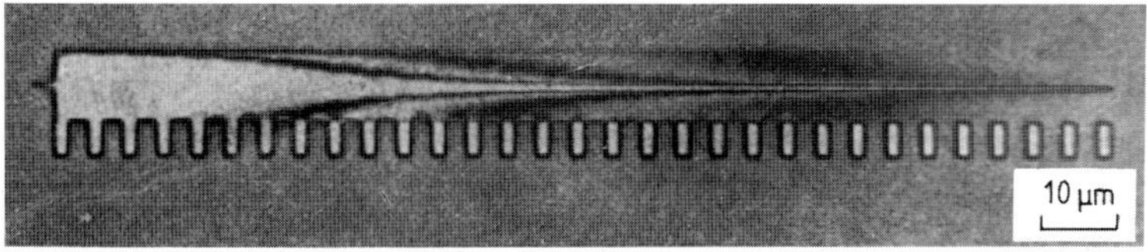

Figure 5.11 Micrograph of the test structure after development [223].

(figure 5.10(*c*), (*d*)). If we use the approximation in which, for high-contrast development, the boundary of the resist corresponds to a line of equal absorbed energy-density W_0, the line describing the boundary of the surface developed down to the substrate will correspond to the distribution described by (5.12). In contrast to the other methods, this procedure determines the entire absorbed energy-distribution function, including its values in the region near the ordinate axis.

It is important to note that α_1, α_2 will be different for different combinations of resist and substrate because they are the result of the combined effect of many factors. Figure 5.11 shows a micrograph of a test structure recorded in PMMA deposited on silicon. The markers along the wedge indicate the exposure steps [223]. By measuring the width of the test structure it is possible to determine the width of the part of the structure developed down to the substrate as a function of the exposure dose.

The fabrication of galvanic zone plates is an example of the application of this method. A silicon wafer coated with nickel, 50 nm thick (the base for the galvanic deposition) and carrying a layer of the EPL-20 resist, 0.6 μm thick, is exposed to a beam of 30 keV electrons (beam current 35 nA). The beam step is 0.125 μm and the exposure time is 528 μs for a point on the strip and 0–600 μs for a point on the wedge. The separation between the markers corresponds to a change in the incident dose by 3.8 μC m^{-2}. Figure 5.12 shows the measured

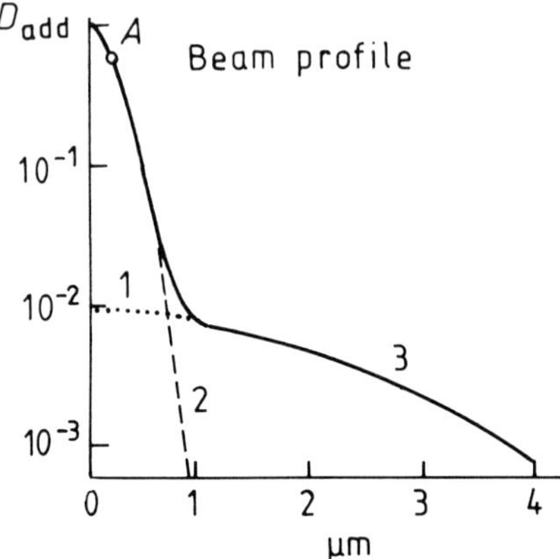

Figure 5.12 Distribution of additional exposure dose D_{add} with distance from the line centre [223]. (1) Back scattering from the substrate, (2) scattering in the resist, (3) total exposure.

half-width of the space between the resist boundaries developed down to the substrate. It was described approximately by the sum of three Gaussians:

$$D_{\text{add}}(x) = \sum_{i=1}^{3} A_i \exp\left(-\frac{x^2}{\alpha_i^2}\right) \tag{5.20}$$

where D_{add} is additional dose necessary for the development of the resist down to the substrate. It is absorbed by the resist as a result of scattering of primary beam electrons:

$$D_{\text{add}} = D_b - D_v \tag{5.21}$$

where D_b is the dose necessary for development down to the substrate, D_v is the variable exposure over the rectangle, which varies according to the graph of figure 5.10(b), and x is the distance from the centre of the exposed line in microns.

The approximation parameters for this particular combination of PMMA and silicon substrate are: $A_1 = 1$, $A_2 = 0.44$, $A_3 = 0.06$, $d_p = 0.21\,\mu$m, $\alpha_1 = 0.33\,\mu$m, $\alpha_2 = 2.7\,\mu$m [225]. The approximating curve is shown by the thick line in figure 5.12. The shape of this curve is very different from the theoretical model discussed at the beginning of this section in which the two-Gaussian approximation (5.9) was used to calculate the dose. The model does not take into account the development process which contributes significantly to the shape of the developed region [226, 227]. From the standpoint of experimental implementation, the function shown in figure 5.12 gives the true experimental parameters of the proximity effect, representing all the phenomena that occur during exposure and processing of the resist. It follows that by substituting it in (5.15) in place of the linear energy dissipation function $\phi_{\text{lin}}(x)$, we obtain a reasonable approximation to the actual experimental condition.

5.4   DEDICATED EBL CONTROL SYSTEM

Commercially available automated systems incorporate dedicated computers, connected to a peripheral processor in the form of a small computer, that process the incoming data and convert them into a signal that controls the position of the coordinate stage and the electron optics.

The software for this configuration must prepare and store the initial data, transform them for the dedicated computer and control the exposure. Most commercial EBL systems make no provision for dose correction, and such systems cannot be used to fabricate nonlinear profiled structures for the complex elements used in X-ray optics. This was the reason why it was necessary to develop a special research system for electron-beam lithography, based on the ZRM-12 system and a microcomputer that combine the functions of the controlling machine and the peripheral processor [236]. A block diagram of the control system is shown in figure 5.3 [235]. Interfacing to the object is accomplished by standard CAMAC modules. The control system consists of four subsystems that respectively perform the following functions: control of the electron-optics, precise positioning, control of the displacement of the stage and interfacing to the control panel. The electron optics is controlled by a dedicated device run by a microcomputer via the CAMAC line.

A dedicated processor is used to perform the exposure of rectangle, strip, point, inclined line at any angle and preset variation of exposure dose (by varying the exposure time in the range $1\,\mu s$–$32\,ms$ in steps of $0.5\,\mu s$.

The controlling microcomputer prapares and stores the incoming data for the control program and processes the data for the dedicated processor that controls the electron optics, the coordinate stage and the exposure. The controlling microcomputer is interfaced to the lithographic part by CAMAC modules.

The software used in the system consists of programs DIALOG, EXPONI and SS whose structure is illustrated in figure 5.13 [235, 236].

DIALOG is the main program used in the fabrication of structures under manual input of coordinates by the operator in the interactive mode. The required topology is introduced in the form of rectangular elements, with the exposure specified for each element. DIALOG can also be used to create and edit a structure file. The program automatically divides the topology into exposure fields, displays the structure file on the graphics monitor and translates the structure file to the format of the control program EXPONI which knows the dimensions of the electron optics.

Program SS translates structure file A into file B which has different geometrical dimensions and exposure dose distribution that compensates for the proximity effects. The program is based on the algorithm described above. The operation of the system relies on the high-level language FORTRAN-4C which means that the software can be constantly upgraded to more powerful computers.

Algorithms for the fabrication of nonlinear structures using polynomial

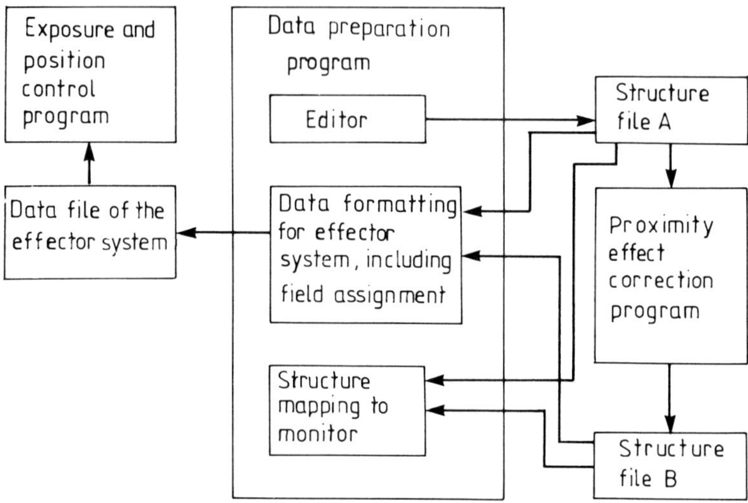

Figure 5.13   Block diagram of software used to control the electron-beam lithography system.

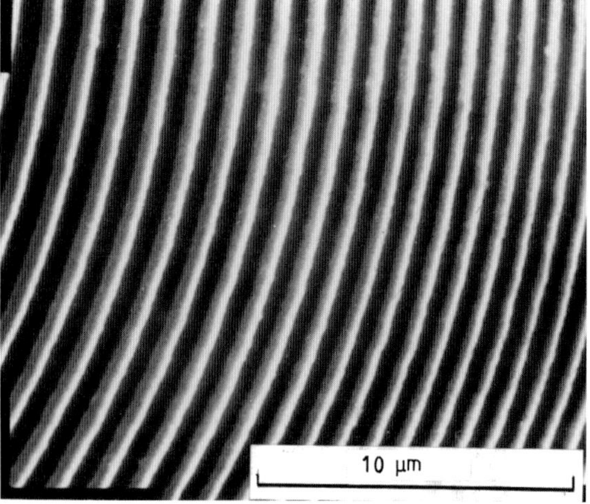

Figure 5.14   Fragment of a zone plate (gold on polyimide membrane).

splines. We have seen that the topology of elements used in diffraction optics is described by second-degree curves. In the simplest case of transmission zone plates, the topology is fabricated in the form of concentric rings with diameters distributed in accordance with a given rule (figure 5.14) [238].

Such systems are often fabricated [201, 207, 239] by incorporating in the scanning system fast real-time translation from Cartesian to polar coordinates.

Such systems are much faster than standard systems for EBL control in the case of circular zone-plate structures. However, they are inconvenient for quasi-elliptical and interference topologies specified in general form, as required in the fabrication of Bragg–Fresnel elements. Moreover, exposure of such topologies with point by point calculations ties up an anormous amount of computer memory and consumes considerable time. The image generator described in [237] produces a broken line that corresponds to the first-degree polynomial spline. This reduces the exposure time for a zone plate of any curvilinear shape by one or two orders of magnitude as compared with direct calculation and point by point exposure.

This interpolation is possible because the EBL system has limited precision and the use of precise data introduces a redundancy. The topology can be approximated by a broken line to a given degree of precision and this has no effect on the final accuracy.

The interpolation of a function by piece-wise splines, which describes a topology of a particular form, is specified with precision e that does not exceed the precision of the writing system [237]. The spline points are defined by a formula that involves e and the function $f(x)$ being interpolated:

$$x_{i-1} = x_i + h\left[e, f(x)\right] \qquad x \in [a, b] \tag{5.22}$$

$$h\left[e, f(x)\right] = 2(2e/\|f(x)\|)^{1/2} \geqslant h_{\min} \tag{5.23}$$

where $[a, b]$ is the interval on which $f(x)$ is specified and $h_{\min}$ is the minimum step (precision) of the writing system. The number of points on the spline grid is given by

$$N = (b - a)(8e)^{-1/2} K\alpha + 1 \tag{5.24}$$

$$K\alpha = \frac{1}{b - a} \int_a^b |f(x)|^{-1/2} \, dx \tag{5.25}$$

and is much smaller than the number of points in the case of direct point by point exposure. For each particular function there is an optimum grid that can be found analytically.

The exposure software reported in [228] for second-degree curves can be used to produce ellipsoidal, parabolic, hyperbolic structures. The spline method is particularly conveninet for structures that can be expressed in canonical form. For example, the following expressions give the spline step size for three second-degree figures:

for the ellipse $\qquad x^2/A^2 + y^2/B^2 = 1 \qquad h_1 = 2\sqrt{2e}\,\dfrac{A}{B^2}|y_1|^{3/2}$

for the hyperbola $\qquad x^2/A^2 - y^2/B^2 = 1 \qquad h_1 = 2\sqrt{2e}\,\dfrac{A}{B^2}|y_1|^{3/2}$ $\qquad$ (5.26)

for the parabola $\qquad y^2 = 2px \qquad h_1 = 2\sqrt{2e}|y_1/p|^{3/2}.$

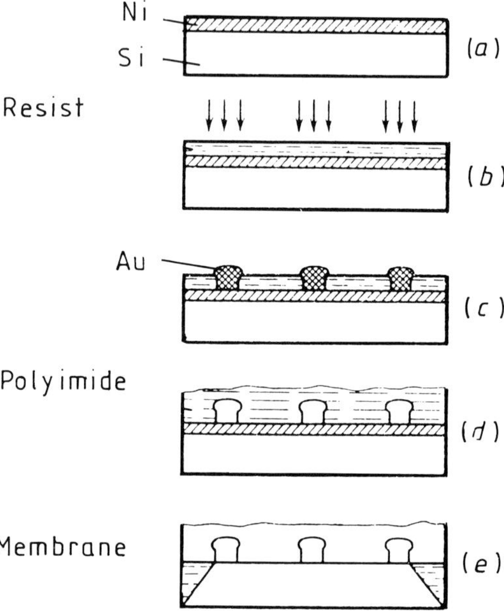

Figure 5.15   The sequence used to produce a zone plate with a gold mask [238]: (*a*) chemical deposition, (*b*) exposure to electron beam, (*c*) electrolytic deposition of gold, (*d*) removal of resist and deposition of polyimide, (*e*) etching out of window in silicon.

The approximation to structures by the spline method can also be used as a basis for image-correcting software. A dedicated control system was used to produce all the experimental structures whose images are reproduced in this monograph.

5.5   FABRICATION TECHNOLOGY FOR ZONE PLATES

Amplitude zone plates. Amplitude zone plates based on gold are currently the most popular and widely used. The properties of Fresnel zone plates were described in detail in chapter 2 and the basic formulae were reproduced there. Figure 5.15 shows a typical fabrication sequence used to produce a zone plate on a polyimide film acting as a substrate. A nickel layer, $0.01\,\mu$m thick, is deposited chemically on a specially prepared (100) or (111) silicon wafer. It is then covered by the electron resist 0.3–$0.5\,\mu$m thick. After exposure to the electron beam in the EBL, the wafer is placed in the developer and inserted into an electrolytic bath. The thickness of the gold film thus deposited on the nickel layer is controlled by the deposition time and the current in the supply circuit.

At the end of the electrochemical deposition process, the wafer is carefully treated with strong organic solvent (dimethylphthalate) to remove any residues of the resist. The cleaned and washed wafer bearing the profiled gold film is then coated in a centrifuge with 3–5 μm of polyimide which is polymerized at 300 K for one hour. The next step is to remove the silicon substrate with a strong acid etching solution. Gold and polyimide are unaffected by the etching solution and all one is left with is the self-supporting membrane with the gold drawing *imprinted* upon it. The basic process in this sequence is lithographic: it produces the submicron topology required for the optical element.

Since the radius of the zones amounts to tens or hundreds of μm and the characteristic dimensions involved in the primary-electron scattering function are of the order of 1 μm, the correction of proximity effects is essentially a linear problem. The main technological parameter is the discreteness of the exposure field. Even if we use a 14-bit DAC, the beam step size on a 1×1 exposure field is 0.0625 μm. A line 0.3 μm thick, is then covered by five or six traverses, so that the dose may vary by as much as 20%. This means that modelling of the final resist profile must include calculation of the distribution of absorbed dose and allowance for the development of the resist for given contrast. Figure 5.14 shows a circular zone plate produced by this type of technology.

Phase zone plates. Zone plates made from gold have low efficiencies ($\sim$5–7%). It was therefore suggested in [70] that it would be better to use nickel and germanium as phase materials for the mask of a zone plate. Such elements are amplitude-phase devices and have efficiencies of 15–17%. For wavelengths of $\sim$0.5–10 nm there are several materials with good phase properties. They include silicon which has good transparency in the range $\lambda = 0.7$–2.0μm (chapter 2). In addition, the microstructuring technology is well established for silicon—the traditional material in microelectronics—which means that elements with linear dimensions of a fraction of a micron can be realistically produced [67].

Fabrication of phase zone plates. The fabrication of Fresnel microzone plates described in [67] involved the basic processes of microelectronics technology such as electron-beam lithography, reactive ion etching, deep doping and anisotropic liquid etching. The two main stages were: deposition on silicon of $\sim 3 \mu m$ thick membranes with a working area of 1 mm^2 and production of freely-hanging masks. A 3 μm stop layer was produced on a boron-doped (100) surface of single-crystal silicon for subsequent anisotropic liquid etching, and a thermal procedure was used to grow an oxide layer, 0.8–0.9 μm thick, over the entire surface. Photolithography and liquid etching on the back of the plate were employed to open up windows in the oxide to receive the membrane. Silicon was then etched anisotropically with a ethylenediamine/pyrocatechin/water solution at 100 °C. At the end of the process, the oxide film on the boundary with the stop layer was removed. A 0.7 μm protective aluminium mask was produced on the resulting membrane by a combination of electron-beam lithography, photolithography (formation of rigid supporting ribs for the ring structures), thermal evaporation and flash lithography. Reactive ion etching of the membrane was performed in a diode-type system, using a high-frequency low-pressure discharge in $SF_2 + O_2$. The plates were cooled during the precision transfer of topological dimensions and the vertical profile etching. The linear and circular self-supporting zone plates produced by these procedures are shown in figure 2.11 [238].

5.6 FABRICATION TECHNOLOGY FOR BRAGG–FRESNEL LENSES

The technology used to fabricate Bragg–Fresnel multilayer lenses involves all the main processes described above. Since these optics components are based on multilayer mirrors, the multilayer Bragg–Fresnel lens substrate preparation is an important step in the fabrication process.

It was found that magnetron sputtering is one of the best processes to deposit amorphous multilayers, especially for metals which have a high electron density but are difficult to evaporate.

The magnetron sputtering system developed in the Laboratory of Electromagnetic Optics (LOE), Marseilles University, and made by the ALCATEL company in 1986, was the first successful prototype of such a system. The next generation of this system fabricated by ALCATEL under LOE consulting was installed at IMT, Chernogolovka (Russia) and in Saint Gobain at Pont E. Musson (France) in 1992–93. A system of the same principal design is now being operated by the CILAS company in Orleans and is commonly used to produce high-reflectivity mirrors with a homogeneity better than 1% along a one meter substrate size.

The first sputtering machine built in LOE is designed to deposit amorphous multilayers with good reproducibility of bilayers and sharp interfaces. The sputtering system consists of two cathodes which are turned on during the whole process (figure 5.16). A rotating substrate holder and a rotating shutter allow the substrate to be exposed to the cathodes. Magnetron sources are used to decouple the plasma from the sample. By this means that any damage the substrate introduced by high-energy ions and secondary electrons is minimized.

Two separate radio-frequency generators are connected to the rectangular cathodes to obtain the best stability in the deposition process. This system is equivalent to two separate chambers where no interactions can occur. Two stepping motors are driven by a microcomputer which controls the positions of the shutter and the substrate holder. The thicknesses are controlled by maintaining the deposition time. For this reason we need perfect stability of all sputtering parameters during the full time of the coating process. In both the first- and the second-generation systems it is possible to perform two types of deposition process: the static process and the dynamic process. In the first one, the substrate is fixed in front of each cathode during each layer deposition time. The second process is performed by one or more deviations with a constant appropriate speed to insure exposure time in relation to the thickness of the layer. This method is generally used to increase the homogeneity of the sample.

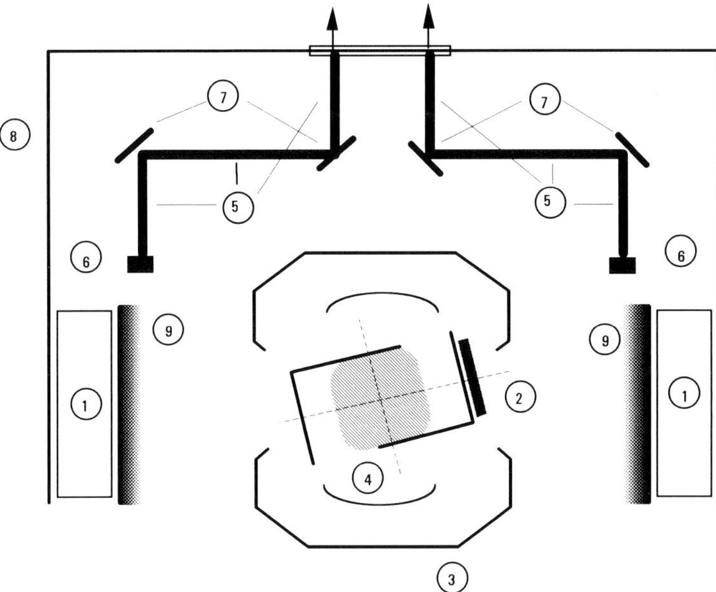

Figure 5.16 Scheme of the magnetron sputtering system designed in LOE, Marseilles University. (1) Magnetrons; (2) sample; (3) shutter; (4) sample holder; (5) periscopes with their (6) micro-channel plates and (7) mirrors; (8) vacuum chamber; (9) argon plasma.

The residual pressure in the vacuum chamber is about 10^{-7} mbar and the argon pressure during the sputtering is 2×10^{-3} mbar which is measured directly by an absolute pressure transducer and controlled by a feedback loop so that any pressure drift during the sputtering is excluded. The typical reflectivity at 8 keV photon energy of the C/W and Si/W multilayers produced by magnetron sputtering is 65–75%.

However, internal stress in carbon tungsten multilayers do not allow a multilayer period thinner than 2.5 nm. This value seems to be the limit due to the reactivity of carbon and the high energy of sputtered carbon atoms arriving at the substrate. Consequently, the amorphous structure of the layer can change to a low-density graphite layer which corresponds to an increase of the period of a multilayer. In order to prevent this it is possible to change the carbon layer to boron carbide. Good results were also obtained with a small amount of boron co-sputtered with the carbon during the deposition of an alloy layer $B_x C_{(1-x)}$.

Several methods can be used to control the layer quality during deposition (chapter 4).

The spectroscopic measurements technique of the glow discharge during the multilayer fabrication was developed at LOE. Two pipes as a periscope with two mirrors, shown in figure 5.16, are used to collect a part of the light emitted by the

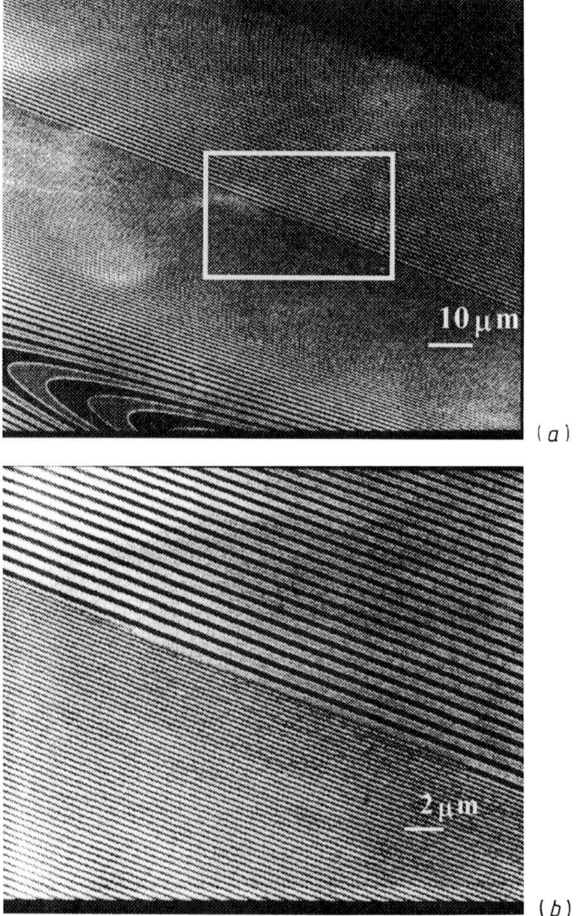

Figure 5.17   Composed multilayer Bragg–Fresnel lens with dimensions of 0.36 mm × 14 mm and minimum zone width of 300 nm.

two cathodes. To prevent deposition on the first mirror of this system a micro-channel plate is used. At the exit of the each pipe is installed a focusing system and optical fibre. They are linked outside the coating chamber at the entrance of a spectrometer. This system has a multi-channel silicon array photodiode as a detector and sends intensity measurements of the glow discharge spectral lines every 25 milliseconds to the monitoring computer.

These measurements are made at the beginning and at the end of the realization of each layer. The ratio between a particular argon line and a characteristic line of the layer material during the layer deposition is computed in real time. The measurements allow verification of the stability of the deposition process. It is than possible to correct immediately the principal parameter of the sputtering process leading to a high degree of reproducibility of the layer thickness and structure.

The fabrication process of the 'composed' Bragg–Fresnel lens is given here as an example.

Electron beam, optical lithography and ion-beam etching processes are employed to obtain a BFML. At the IMT laboratory a ZRM-12 electron-beam instrument with a dedicated computer control system has been employed for the generation of an elliptical-shaped topology of a BFML. The hardware as well the software for the IMT exposure system was developed to provide topology fabrication of complex optical elements. An RF-excited argon ion-beam source for etching the BFML structure in the multilayer was used.

To avoid scattering from the mirror around the BFML the second lithography-etching procedure was performed through the special aperture mask. The aperture mask was patterned in a layer of Fe_2O_3 on a glass substrate by electron beam lithography and wet etching. This mask was used for UV lithography on the multilayer coated with photoresist. Finally, the multilayer was etched down to the depth of 170 nm through the photoresist mask by using the same RF-excited ion-beam source.

In the same substrate several lenses were prepared in order to compare their properties: one with only first order and another composed-type design. Two pairs of lenses were made: for 8 keV and 13.8 keV radiation. One of the composed lenses is shown in figure 5.17. A zone structure was calculated using the equation (2.28(a), (b)). Lenses were prepared on a 60-bilayer, 2.98 nm period W/C multilayer mirror coated by magnetron sputtering. An magnetron coating system, designed with the help of LOE, Marseilles, France was used for the multilayer preparation.

6

Diffraction optics applications: X-ray microscopy

There is growing need in every human field of activity for the determination and characterization of elements at trace concentration, that is below one part per million by weight. For qualitative as well as quantitative analysis X-ray microanalysis is often used as a signature of elemental composition. This first step of characterization is becoming insufficient and a map of the distribution of each element is more and more desired with the highest lateral resolution. This can be reached with existing X-ray microscopes or microprobes, the great majority of which use electrons or high-energy photons as projectiles. Only very recently, by using diffraction X-ray optics, have photon microprobes become available, such as the one at LURE (Orsay, France) which we present here in detail. One can find below a description of the three basic methods used in X-ray microscopy. All of them have been realized or are on the way to realization using diffraction X-ray optical focusing elements.

6.1 PHYSICAL PRINCIPLES

The absorption of X-ray photons by inner-shell electrons is the basis of X-ray absorption spectroscopy (XAS), X-ray photoelectron spectroscopy (XPS) and X-ray fluorescence analysis (XFA). It results in the weakening of the corresponding spectral lines in the sample, the excitation of photo- and Auger electrons and the emission of characteristic X-ray fluorescence [240]. These processes provide information on the elemental and chemical composition of the sample under investigation. Figure 6.1 illustrates the processes that take place in an atom when it interacts with X-ray photons.

The basic equation of photoelectron spectroscopy is

$$h\nu = W_c + W_k \tag{6.1}$$

where $h\nu$ is the energy of the X-ray photon, W_c is the binding energy of an atomic electron and W_k is the kinetic energy of a photoelectron.

The kinetic energy of ejected inner-shell electrons (photoelectrons) is proportional to the energy of the incident photons. When the incident photons

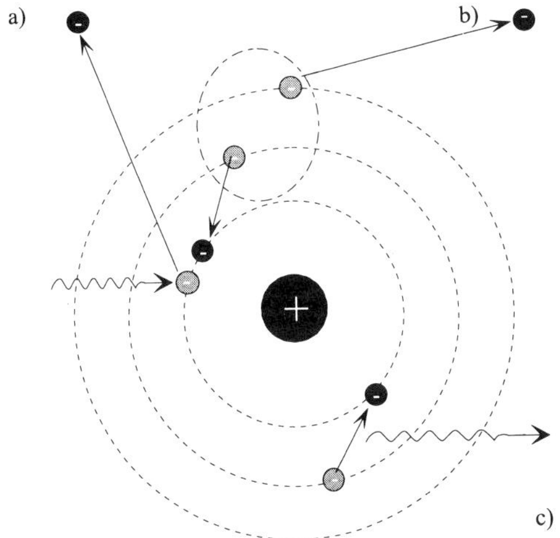

Figure 6.1 Secondary processes in an atom interacting with X-ray photons: (*a*) photoelectron emission, (*b*) Auger–electron emission, (*c*) fluorescence.

are accurately monochromatic, measurement of photoelectron kinetic energy provides quantitative information on the binding energy of electrons in the atom and, hence, on the elemental and chemical composition of the sample. The accuracy with which the position of an emission peak can be determined depends on the width of the radiation line, e.g., when normal sources of soft X-rays are employed, the corresponding figures are 0.7 eV for $K\alpha(Mg)$ and 0.85 eV for $K\alpha(Al)$. If we assume that the line energies are, respectively, 1253.6 eV and 1486.6 eV, the limiting resolution is approximately 6×10^{-4}. To improve the resolution, the relative resolution of the analyser must be better than 10^{-4}, which is very difficult to achieve in practice. Analysers different from those employed in Auger spectroscopy (XAS) are therefore used to record XPS spectra.

The photoelectrons are usually slowed down to a particular energy, i.e., the energy of analyser transmission. When the electrons are slowed down to a transmission energy of 50 eV, an absolute resolution of 0.7 eV then requires a relative resolution as low as 0.01 eV. A retarding field can be used to obtain the same absolute resolution at rather poorer relative resolution [241].

The ejection of an electron from an inner shell W_x of an atom (figure 6.1) produces a vacancy that is filled by an electron from a shell W_y as a result of internal processes in the atom. The energy $W_x - W_y$ released in this way is carried by the electron from the shell W_z. Under photon excitation, this process is naturally less probable than photoelectron emission. The basic equation of the Auger process induced by an incident X-ray is

$$W_k = W_x - W_y - W_z. \tag{6.2}$$

The energy of the Auger electrons does not depend on the energy of the incident X-ray photons. The optimum photon energy must exceed the energy of the Auger peak by a factor of three or four.

X-ray fluorescence (XFA) is the fourth process accompanying the interaction of X-rays with atoms. Just as in the Auger case, a vacancy in an inner shell is filled by n electrons from one of the outer shells, and an X-ray photon is emitted with energy characteristic of the given element or bond (see [242]). This process competes with XAES, but is less probable. The basic equation for X-ray fluorescence is

$$h\nu_f = W_x - W_y \tag{6.3}$$

which gives the position of a characteristic line of the given sample.

X-ray acoustic microscopy (XAM) is among possible methods of X-ray investigation that are now under development [243]. XAM can be used to examine samples with three-dimensional resolution. The acoustic signal yields not only the composition of the sample surface, but also the depth at which an inhomogeneity occurs. Unfortunately, the depth resolution is limited by the wavelength of the acoustic wave, and is no better than a few micrometres.

The actual yield of the interaction between a sample and X-ray photons is discussed below.

6.1.1 X-ray absorption spectroscopy

Within the energy range corresponding to soft X-rays (say, above 50 eV), only coherent scattering and photoabsorption need be considered (see section 1.1).

The component f_2 of atomic scattering is directly propotional to the photoabsorption cross section whilst f_1 is determined by the dispersive integral which is also directly proportional to the photoabsorption cross section. It is therefore important to know whether the photoabsorption cross section depends on the state of the target substance or whether it can be determined as the sum of its component cross sections. This problem is discussed in the literature which indicates that, at energies above 50 eV, the absorption spectra behave as if the substance consisted of isolated atoms. For energies below 50 eV, valence bonds begin to play an appreciable role and the photoabsorption cross section depends on the state of the system. The role of the imaginary and real parts of the scattering factor is discussed in more detail in chapter 1. It should be noted that the phase contrast of the target substance is beginning to be exploited in the latest techniques of X-ray microscopy as a source of further information on the composition and structure of an object. The components f_1 and f_2 are determined in chapter 1 (equations (1.1) and (1.3)).

The monatomic absorption factor (μ_a) can readily be converted to the mass and linear absorption factors (μ_g and μ_t, respectively) by using the following expressions:

$$\mu_a \left[\frac{\text{cm}^2}{\text{atom}} \right] = \frac{G}{N_a} \mu_g \left[\frac{\text{cm}^2}{\text{g}} \right] = \frac{G}{N_a \rho} \mu_t \left[\text{cm}^{-1} \right] \tag{6.4}$$

where G is the atomic weight, ρ is the sample density and N_a is the Avogadro number.

Let us now estimate, for an X-ray microprobe, the efficacy of photoabsorption spectroscopy in the local analysis of elemental composition of a sample of nanometre thickness. The intensity of radiation transmitted by the absorbing medium is given by the standard expression

$$I_1 = I_0 \exp\left[-\sum_g \mu_a(g)N_g t\right] \tag{6.5}$$

where N_g is the concentration of atoms of type g (atoms per cm^3) and t is the thickness of the layer under investigation (cm).

The sum in the exponential represents an approximation in which the photoabsorption cross section does not depend on the state of the substance being analysed. We shall suppose that the element under analysis is distributed in a film of thickness t and the quantitity measured is the image contrast obtained when the probe of area S_z scans its surface.

Microparticles of one material, imbedded in a host of a different material, are investigated most commonly. The minimum detectable film thickness of the host and the minimum detectable impurity concentration may then serve as the parameters for estimating the efficacy of the method.

Suppose that the host has an atomic cross section $\mu_a(2)$ and atomic concentration N_2, and that the impurity atoms have the absorption cross section $\mu_a(1)$ and atomic concentration N_1. The image contrast is then given by

$$K = \frac{1 - \exp\{-t\,[\mu_a(2)N_2 - \mu_a(1)N_1]\}}{1 + \exp\{-t\,[\mu_a(2)N_2 - \mu_a(1)N_1]\}}. \tag{6.6}$$

If the optimum image contrast is 10%, the minimum detectable concentration (the detection limit) is

$$N_1 = \mu_a^{-1}(1)\left[0.2/t + \mu_a(2)N_2\right]. \tag{6.7}$$

This formula can be used to estimate the minimum detectable thickness of the material under investigation (in μm). The result is

$$t \geqslant 2 \times 10^{-11}/\mu_a(1). \tag{6.8}$$

Since the photoabsorption cross section in the wavelength range 1–10 nm is 10^{-10}–10^{-11} μm per atom, the minimum detectable thickness of a pure homogeneous material lies in the range 0.1–0.01 μm.

XAS is used extensively with samples of light elements, especially biological objects, and soft X-rays. X-ray microscopes now rely on synchrotron sources with high-quality radiation, i.e., high spectral brightness, small divergence, small source dimensions and so on.

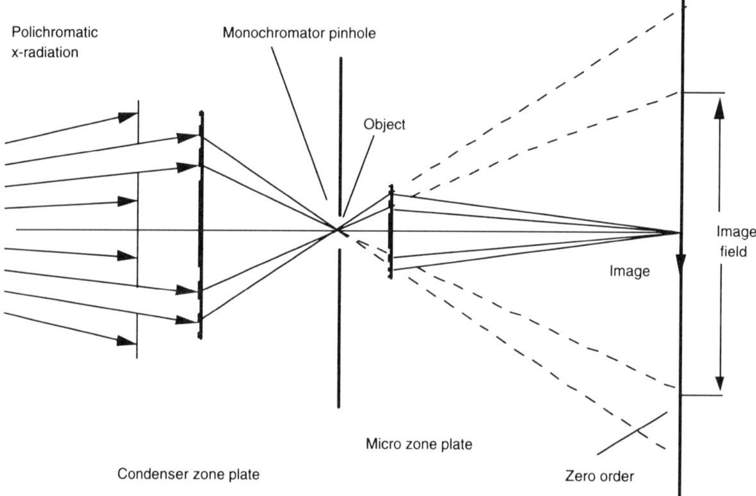

Figure 6.2   Optical system of the X-ray imaging microscope incorporating zone plates.

The best source for X-ray microscopy is the *undulator* (chapter 3) which ensures high spatial resolution.

XAS microscopes can be divided into imaging and scanning. In the former (figure 6.2), the optical system produces a $\times 10$–30 magnification of the object area illuminated by X-rays from the condenser lens. The image is recorded either on photographic film or by a photoelectric detector. Diffraction zone plates are used as condenser and imaging lenses. The spatial resolution of such microscopes can reach 75 nm with solid state samples [244] and 50 nm with biological samples [252].

XAS microscopy using synchrotron radiation [244–249] and laser-induced plasma sources [250, 251] is being developed in several centres. X-ray absorption spectroscopy (XAS) has been used to investigate biological objects by visualizing the elemental composition of a sample in the scanning regime. The image contrast is significantly altered when scans are made within the range of absorption edges of the main elements of biological objects, i.e., nitrogen, carbon and oxygen, thus elucidating the components of elementary compounds.

Imaging of an object requires special recording methods (multichannel optic analyser, photodetector matrices, *etc.*). Moreover, the distortion of the image field influences the resolution. The X-ray optical system that produces the sharply focused beam in scanning microscopy is an alternative to the XAS system. The sample is scanned by mechanical or piezoelectric displacement in the plane of focusing. Several microscopes of this design are being successfully used in the investigation of biological objects and microelectronic devices [244–249]. The scanning microscope (figure 6.4) employs a fairly simple recording system with graphical display.

X-ray absorption microscopy is used to investigate transparent materials prepared as thin membranes. The analysis of thick films or bulk samples can be based on the above-mentioned secondary effects such as X-ray photoelectron spectroscopy (XPS), X-ray fluorescence analysis (XFA) and so on.

6.1.2 X-ray local photoelectron spectroscopy

The sensitivity of XPS is determined by the photoelectron flux required to record a signal. A simple relationship can be used to estimate the limiting sensitivity of this method [242]:

$$I_{\mathrm{phe}} = \Phi(\lambda)\beta_{\mathrm{phe}}(g, \lambda)Q \qquad (6.9)$$

where I_{phe} is the flux density of the photoelectron signal (electrons per $s\,cm^2$), $\Phi(\lambda)$, $\beta_{\mathrm{phe}}(g, \lambda)$ is the quantum yield of photoemitting atoms of type g and Q is the photoelectron collection efficiency of the detector. The photoemission quantum yield can be estimated from the expression

$$\beta_{\mathrm{phe}}(g, \lambda) = N_g \mu_a(g, \lambda)t \qquad (6.10)$$

where N_g is the number of atoms of type g, $\mu_a(g, \lambda)$ is the photoionization cross section at the excitation wavelength λ (cm^2 per atom) and t is the depth of emission of photoelectrons from the surface. For photoelectron energies between 50 eV and 2 keV, we find that $t = 2.5 \times 10^{-3}\,\mu m$.

We shall now consider estimates of the limiting sensitivity and resolution of the method. The photoelectric ionization cross sections of the substances for 1.5 keV X-rays lie in the range 10^{-18}–$10^{-19}\,cm^2$ per atom. We note that the photoelectric ionization cross section is a maximum near the ionization threshold [242]. At least 10^4 photoelectrons per image element are needed to obtain contrast of about 10%. Since the detector efficiency is of the order of 1%, the total number of photoelectrons escaping from an image element must be at least 10^6–10^7.

Let us now estimate the number of X-ray photons necessary to record images with a resolution of 1 μm for an object containing impurities with concentration N_g (atoms per cm^2). From (6.9),

$$\phi(\lambda) = 2.5 \times 10^{31}/N_g. \qquad (6.11)$$

Thus, to determine the composition of pure materials with atomic concentration of about 10^{22} atoms per cm^3, we need at least 2.5×10^9 photons per image pixel. Using equation (6.11), we can readily estimate the detection limit (the concentration) for the impurity in a given X-ray flux. For example, if the X-ray flux is 10^{14} photons per second and the detection time is 10 seconds per pixel, the minimum detectable concentration is 2.5×10^{-16} atoms per cm^3 or 3.5 ppm.

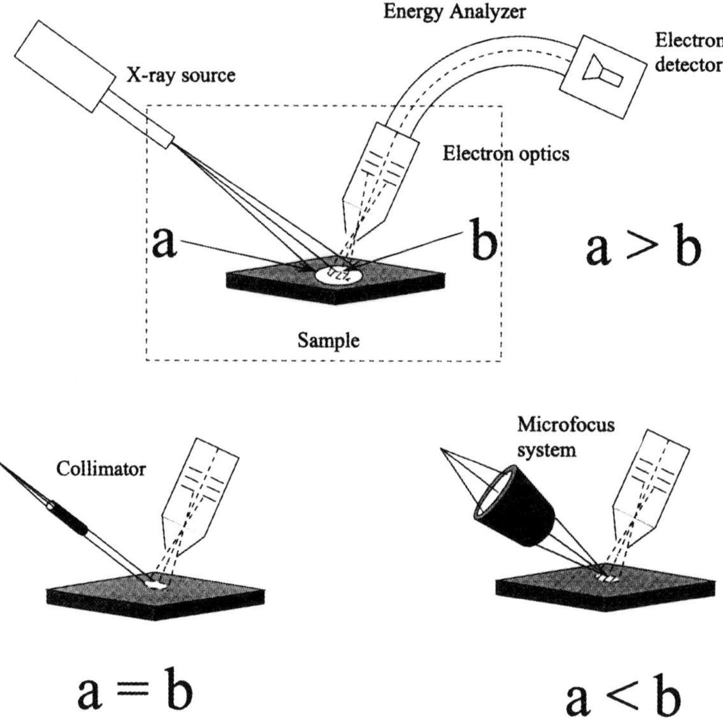

Figure 6.3   Three cases of localization of photoelectrons.

The practical limit of spatial resolution is determined exclusively by the sharpness of X-ray beam focusing and may reach 10 nm when zone plates are employed.

Local photoelectron analysis may be performed in a number of ways that differ by the relative size of the excitation region, the X-ray flux and the region from which photoelectrons are collected (figure 6.3; a is the diameter of the region illuminated by the X-ray beam and b is the diameter of the region from which photoelectrons are collected).

Three special cases may be considered, namely, (1) $a > b$ (limited area of photoelectron collection), (2) $a = b$ (complete image method) and (3) $a < b$ (microprobe method, scanning XPS).

Local XPS was first considered in [253] and the results were obtained two years later [254]. The method consists of sweeping a focused electron beam over the surface of an aluminium film, thus generating characteristic radiation. A thin sample placed behing the aluminium film is illuminated by a point source of X-rays and produces the photelectrons that are collected by a detector. The resolution achieved by this method can reach 10–$30\,\mu m$. However, it is suitable only for objects in the form of thin films.

Table 6.1

	AES (20 keV)	XPS (1 keV)
Minimum particle flux	10^{10} electrons per pixel	10^9 photons per pixel
Ionization cross section	$10^{-18}\,cm^2$ per atom	$10^{-19}\,cm^2$ per atom
Practical resolution limit	50 nm	10 nm
Signal-to-noise ratio	0.1–1	1–100
Radiation load on sample	$160\,W\,mm^{-2}$	$1.6\,W\,mm^{-2}$

The method relying on the localization of photoelectron collection can be illustrated by the device described in [255], which is now typical of commercial systems [256]. When conventional X-ray sources are used, the spatial resolution can be up to 0.2 mm. The sample is illuminated by an erect X-ray beam over the whole area. The area under analysis is limited by the field of view of the transmitting electron lens. The emitted electrons are slowed down and directed into an electron spectrometer. A multichannel detector can be used in the system.

Powerful sources of X-ray radiation such as electron synchrotrons and storage rings enhance the high-resolution photoelectron image . The microscope described in [257] is an attempt to exploit this feature. A sample is placed on the surface of a film detector and is illuminated by an X-ray beam. The photoelectron image produced on the other side is enlarged by an electron optical system and is then detected. This system is actually the X-ray absorption microscope (XRAM) with photoelectron detection.

The system MICRO ESCAS developed by the *Surface Science Instruments* company for the Stanford Synchrotron Radiation Laboratory is an excellent X-ray local photoelectron spectrometer for the chemical analysis of sample surfaces. A superconducting solenoid surrounding the sample forms the bulk of this system. The magnetic field (7 T) established by the solenoid produces an enlarged photoelectron image of the excited area on the detector located at 2 m from the surface. Images have been obtained with a resolution of up to $5\,\mu$m [258]. The scanning microprobe method in which the X-ray beam beam is focused in a small spot is very difficult to implement. Several designs have been published for such devices [259, 260]. The main problem is the development of an X-ray optical system capable of producing a sharply focused X-ray probe.

Local XPS is of particular interest for the analysis of surfaces. Comparison of this method with local Auger electron spectroscopy (AES) using excitation by an electron beam shows that XPS is not only comparable with AES for standard conductors, but is also suitable for the analysis of dielectrics and even X-ray sensitive materials. The parameters of the two methods are compared in table 6.1.

The absence of secondary electrons and, hence, the high signal-to-noise ratio, and also the shorter analysis time at each point, are among the principal

Table 6.2

Objects	Spatial resolution μm
Polycrystals	10–100
Alloys, grain boundaries	0.1–10
Magnetic powders	0.05–10
Electronic devices	0.01–1
Clusters	0.001–0.05

advantages of XPS as compared with AES. The targets examined by XPS and the necessary spatial resolutions are listed in table 6.2.

We note that the examination of electronic devices and cluster systems requires resolutions close to the practical resolution limit of X-ray optics.

6.1.3   X-ray fluorescence analysis

Whilst local fluorescence analysis (XFA) is suitable only for relatively thick samples, it is capable of submicrometre transverse resolution. Moreover, beam spreading in thick samples is very small because of the small X-ray scattering cross section, so that the X-microprobe is found to have the best spatial resolution of all the known methods of fluorescence excitation, including VUV and optical techniques.

Estimates based on diffraction spreading of the beam and the depth of penetration show that the actual resolution limit of the X-ray fluorescence microscope is about 50 nm [261]. Taking into account the fluorescence quantum yield and the signal-to-noise ratio, it is concluded that an X-ray microprobe with a spot diameter of between 50 nm and 1 μm is preferable to other forms of microscopy for most samples.

The sensitivities of both XFA and XPS are determined by the minimum detectable flux of X-ray fluorescence required for signal detection. The fluorescence intensity can be calculated by analogy with (6.19)–(6.11). The relevant expression is

$$I_f = \Phi(\lambda)\beta_f(g, \lambda)Q \tag{6.12}$$

where I_f is the fluorescence flux density (photon per $s\,cm^2$), $\beta_f(g, \lambda)$ is the fluorescence quantum yield of atoms of type g (photon per photon) and Q is the fluorescence collection efficiency of the detector.

The fluorescence detector efficiency turns out to be much smaller as compared with XPS because it is impossible to use an extracting field when detecting the photoelectrons. The Si(Li) energy dispersive detector with window diameter of 5–30 mm, placed at a distance of 2–5 cm from the sample, is generally used in fluorescence analysis. It is set at an angle of 90° to the incident radiation in order to minimize noise. An effective value of about 10^{-3} to 10^{-4} is typical for X-ray fluorescence collection by this type of detector.

By analogy with equation (6.10), the fluorescence quantum yield can be estimated from

$$\beta_f(g, \lambda) = N(g)\mu_f(g, \lambda)t \tag{6.13}$$

where $\mu_f(g, \lambda)$ is the fluorescence cross section.

Because of the considerable penetration depth of X-rays in the sample, the output depth of the fluorescence signal is about $100\,\mu$m. This means that, when $\beta_f(g, \lambda)$ is estimated for thin samples, the true thickness should be considered. To examine the true possibilities of the method, we use the data presented in [261] on the fluorescence cross sections of the elements in the periodic table. It is found that the fluorescence cross section usually lies in the range 5×10^{-21}– $10^{-20}\,\text{cm}^2$ per atom for the K and L characteristic radiation of the elements. The corresponding fluorescence yields of pure elements and sample thicker than $10\,\mu$m is then 0.1 for the absorption K-edge and 0.01 for the L-edge. These values exceed by a factor of 10–200 the value obtained under electron excitation of fluorescence.

The signal-to-noise ratio is a further important parameter of the fluorescence signal. Under excitation by X-rays, it amounts to 10–100, which is higher by a factor of 10^4 than for electron excitation.

The above data lead to the conclusion that the number of X-ray photons is smaller by a factor of 10^3 than the number of electrons needed to record a signal providing the same information.

Taking into account the signal-to-noise ratio of the fluorescence signal, we find that 100 photons per pixel must be recorded to obtain good image contrast [261]. By considering the detector efficiency, it is estimated that the threshold number of fluorescence photons during the recording time is about 10^5. Consequently, the number $\Phi(\lambda)$ of primary-beam photons needed to record an image element of the sample containing a given impurity concentration N_g is

$$\Phi(\lambda) = 10^{29}/N_g. \tag{6.14}$$

Thus, at least 10^7 photons per pixel must be focused in the incident beam in any determination of the elemental composition of a pure substance. Comparison with the analogous threshold flux in XPS then shows that fluorescence spectroscopy is more sensitive and can be used to determine concentrations N_g in the range 10^{-13}–10^{-14} atoms per cm^3 s with undulator radiation (1–10 ppb), whereas XFA is suitable only for the analysis of relatively thick samples with vertical boundaries of impurity distribution.

6.2 PRACTICAL SCANNING FLUORESCENCE X-RAY MICROPROBE

As mentioned before, scanning microscopy was first proposed as a way to circumvent the difficulties of X-ray optics [247]. There are several designs of scanning microscopes that employ the microfocus X-ray tube or the transmission X-ray source with scanning electron beam [242]. However, the first successful implementation of this technique for specimen mapping had to await the availability of synchrotron sources [262].

A pinhole collimator with a diameter of 1–2 μm was employed and the fluorescence signal from the specimen was used to map the distribution of different elements in the sample. There are two problems with the application of the pinhole collimator to X-ray microprobe analysis. The first is the low flux transmitted by the pinhole and the second is the small gap between the pinhole and the specimen, so that it is difficult to place the detector in its proper position.

Focusing mirrors have been used to overcome these problems in the NSLS X-ray microprobe. It can be operated with two different optical systems, namely, a pinhole collimator and an 8:1 ellipsoidal focusing mirror [263]. In the first mode, the beam size is typically $8 \times 8\,\mu$m and the flux is about 10^{10} photons per second within an energy range 4–30 keV. The detection limit is generally in the range 1–5 ppm for thick samples.

The Kirckpatrick–Baez microprobe is based on a pair of small spherical multilayer mirrors [264]. It was built at the Centre for X-ray Optics, LBL. Compared to grazing incidence optics, the use of multilayer coatings reduces the mirror size because a larger angle of incidence can be employed. The calculated beam size should be of the order of $3 \times 1\,\mu$m^2 in the X–26C beam line at NSLS, but spherical aberrations lead to a measured beam size of $6 \times 6\,\mu$m^2. The energy of the microprobe is tunable in the range 6–14 keV. Recently a 1.5 μm focal spot diameter was achieved using elliptical bent multilayer mirrors. An *ellipsoidal bent Si crystal* is used in the Daresbury microprobe [265]. The beam size at 15 keV is $10 \times 20\,\mu$m^2 .

A tapered glass capillary was used successfully for condensing X-rays through total multireflection from the inside wall of tapered glass tubes. The great disadvantage of this system is that the beam spreads out very rapidly outside the capillary and the announced resolution is only obtained a few mm from the end of the tube.

Diffraction optical elements must be used to obtain resolutions better than 1–2 μm.

Zone plates with very fine spatial resolution have low efficiency and poor mechanical stability.

Bragg–Fresnel optical elements are good candidates for the microprobe focusing system. The microprobe is based on the elliptical multilayer Bragg–Fresnel lens built at LURE, France in co-operation with IMT of the Russian Academy of Science. The measured beam size is 0.84 μm [266, 267]. The same design but with a spherical Ge Bragg–Fresnel lens was tested at ESRF, France. The beam size is 0.6 μm at 13.5 keV [268].

6.2.1 The LURE–IMT fluorescence microprobe

A description of this installation is given in the book as an example of a high-resolution X-ray microprobe. It is based on the Bragg–Fresnel optical element and has a unique performance regarding spatial resolution and sensitivity especially with a third-generation storage ring such as the European Synchrotron Radiation Facility (ESRF).

In this new kind of X-ray optical system, a multilayer acts as the Bragg reflector and ensures monochromatization of the white incident X-ray beam. The Fresnel zone structure, made by the electron beam lithography and ion etching technique (see chapter 5) on the multilayer mirror provides diffraction focusing of the reflected X-ray beam. In principle BFML keeps the advantage of the Fresnel zone plates which offers high spatial resolution when used in transmission. Moreover they present high mechanical stability as demonstrated with the use of multilayer X-ray mirrors. Such a high mechanical stability is of primary importance considering the photon flux they will undergo with the advance of ultrabright synchrotron sources.

Kirkpatrick–Baez BFML diffraction microprobe
The first test of BFML with synchrotron radiation was done in the soft X-ray range at Siberia-1 SR facility in Moscow in 1989 [166] and in the hard X-ray range at LURE in 1991 [266]. This first kind of photon microprobe used an arrangement of two linear BFMLs mounted on two independent goniometers in the Kirkpatrick–Baez focusing scheme. Alignment of the microprobe was rather time consuming but the results were very encouraging since it was possible to image a 100 mm diameter X-ray source on a $2 \times 3\,\mu\mathrm{m}^2$ focal spot.

Elliptical BFML microprobe
By that time theoretical considerations led to the conclusion that BFMLs with elliptical zone structures would provide better two-dimensional focusing. Such lenses were made at the IMT RAS and tested in LURE in 1992 [269]. This kind of lens is now in operation on the LURE photon microprobe.

The engraving of the lens was done on a 60-bilayer 3.2 nm period W/Si multilayer mirror coated by magnetron sputtering on a thick highly polished silicon substrate. An ALCATEL SCM-651 coating system was used for the multilayer preparation (see section 5.2 and [270–273]). Lenses used now are prepared as phase lenses, the optimum profile depth corresponding to the maximum first-order diffraction efficiency being calculated with the formula for phase conical multilayer gratings (4.69–4.70). For a multilayer period of 3.2 nm, $t_\mathrm{W} = 1.0$ nm and $t_\mathrm{Si} = 2.2$ nm we have $t_\mathrm{opt} = 83.2$ nm (or 28 bilayers). The optical constants of the materials were taken from the Henke tables (see chapter 1). For tungsten $\delta = 4.57 \times 10^{-5}$, $\beta = 4 \times 10^{-6}$; for silicon $\delta = 7.56 \times 10^{-6}$ and $\beta = 1.7 \times 10^{-7}$. The efficiency of large-period gratings can be described by a simple formula that corresponds to the amplitude–phase grating (see chapter 2). Using the χ value:

Table 6.3

Wavelength nm	Dimensions	Multilayer period nm	Minimum zone width nm	Focal length cm
0.1	59 μm × 2.48 mm	3.25	250	15

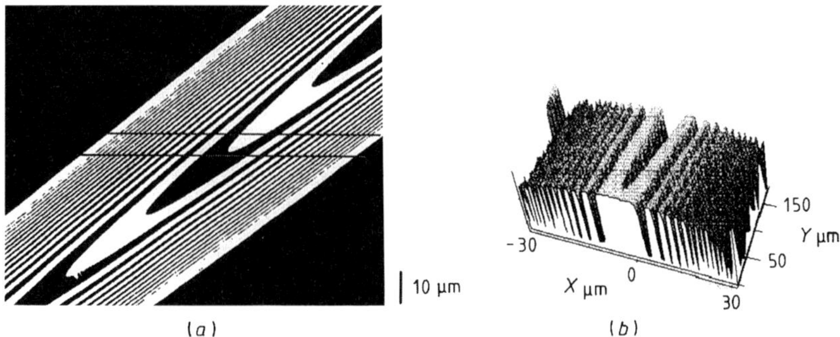

Figure 6.4 Microphotograph of an elliptical Bragg–Fresnel multilayer lens. (a) general view, (b) cross-section.

$$\chi = \frac{\beta_W t_W + \beta_{Si} t_{Si}}{\delta_W t_W + \delta_{Si} t_{Si}} \qquad (6.15)$$

we find for the above multilayer mirrors that $\chi \sim 0.22$ and $E_1/E_0 \sim 0.26$. The main characteristics of the lens used here are given in table 6.3 and a microphotograph of this lens is shown in figure 6.5. Grooved parts of the lens appear as light areas and the multilayer mirror around the lens is etched through a photomask, made by optical lithography.

The diffraction aberration characteristics of this kind of optics can be calculated using aberration formulas (4.49a, b). It then appears that due to the technological limitation in the number of Fresnel zones that can actually be drawn these aberrations are not to be taken into account. The theoretical resolution limit is then mainly fixed by the minimum zone width of the lens. In practice it is the magnification of the lens that will contribute to the focal spot size in our installation.

A drawback of this type of lens is that the focal spot will be surrounded by a halo due to undiffracted zero-order photons. It was measured that there are about as many photons in this halo as in the focal spot. Its contribution can be reduced by positioning a slit after the lens, as far from the lens as possible, but there are then difficulties in setting the detector and the microscope near the sample.

A very interesting solution has been recently proposed and tested using off-axis Bragg–Fresnel multilayer lenses [274]. In this case the focal spot and the

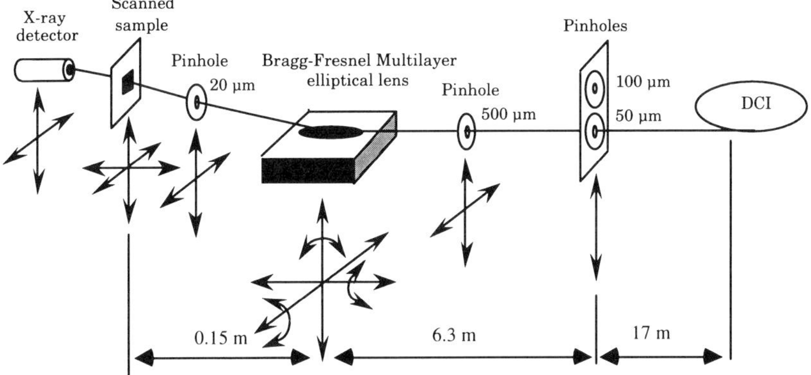

Figure 6.5   Experimental scheme of the LURE-IMT photon microprobe.

zero-order undiffracted beam are separated in space and the latter can be very easily stopped with an absorber placed before the sample.

Optical scheme

The present design of the X-ray fluorescence photon microprobe uses a single elliptical Bragg–Fresnel lens made in IMT RAS (Russia). A scheme of the experimental set up and a photograph of the microprobe are shown figure 6.5. It was installed on line D15 of DCI storage ring (Orsay, France) in July, 1992.

Because of the large size of the electron beam in this first-generation machine, optional entrance pinholes (50 or 100 μm diameter) were added 6.3 m upstream from the lens to simulate a smaller apparent X-ray source. After the beryllium window that isolates the X-ray beam line under vacuum from the rest of the experiment which is under air, a second pinhole (200 μm diameter) is installed to eliminate most of the scattered photons.

The elliptical BFML position is adjusted in the direct beam by a five-axis multi-stage goniometer ($y, z, \theta_x, \theta_y, \theta_z$). To further restrict the scattered beam and cut down most of the undiffracted zero order, a 20 μm diameter pinhole is positioned 10 cm after the lens on a two-axis stage (y, z). Another six-axis stage is available and can be used to position a secondary mirror after the lens if needed (surface acoustic waves experiments or double-reflection system). In fact it was initially needed when the Kirkpatrick–Baez scheme was used with two linear BFMLs. This microprobe was built by RADIKON (St Petersburg, Russia). All of the 13 motors (5 μm step resolution) are remote controlled through a single RS 232 connection with a PC computer.

The sample is positioned in the image plane within an accuracy of 0.1 μm and adjusted at the focal distance within 1 μm steps by another three-axis Micro-Control remote-controlled multi-stage. This stage can be set perpendicularly to the beam for transmission tests or at 45° for fluorescence mapping.

The positioning of the lens is most critical to obtain the best focal spot and maximum intensity. The lens must be set perfectly in the orbital plane of the storage ring for maximum intensity and to take full advantage of the high degree of horizontal polarization in this plane. The short axis of the lens should also be perfectly aligned in this plane and the long axis exactly in the direction of the beam (better than $0.1°$) to reduce aberrations. Finally, the lens engraving being calculated for a given wavelength, the Bragg angle must be set properly. To perform an easy pre-alignment a small He/Ne red laser is used to simulate the direction of X-ray beam in visible light. Then an X-ray camera with 15 μm lateral resolution is used to perfect the setting but the final adjustment can only be achieved by observation of the focal spot with high resolution holographic films.

Another important thing is to correctly place the sample in the focal spot. First a rough position can be estimated using the laser beam. A cross made of two thin nickel wires is then used to record the horizontal and vertical co-ordinate of the spot by measuring the fluorescence intensity while moving the cross in the two directions. When the cross is set on the focal spot, a video colour microscope (magnification $700\times$) is used and two perpendicular reticules are moved and fixed to materialize this position on a colour TV screen. Then the part of sample to be analysed can be moved so as to overlap the reticule crossing.

An image of the screen can be obtained through a video-printer and saved as a visible image of the sample to be superimposed to the X-ray reconstructed images. This step seems complicated and the procedure rather obsolete but due to the low brilliance of the X-ray source of DCI it is impossible to simply observe the impact of the beam on a visible fluorescence screen converter without an intensified camera.

Finally the X-ray fluorescence spectrum is recorded with a Si(Li) detector of 150 eV energy resolution and 13 mm^2 area. An aluminium collimator is installed on the detectors axis for shielding and to limit the acceptance angle. Scattering in the detector is further reduced by placing it in the plane of the synchrotron ring and perpendicularly to the incoming beam. Pulses from the preamplifier are sent to an amplifier and then processed through a Nucleus Multichannel Analyser card set in an IBM PC computer. When working in transmission mode a Si avalanche diode detector of 1.5 mm^2 area and operating at room temperature is used. In the fluorescence mode and with thin samples this detector is also useful for monitoring the transmitted beam.

Special software developed (PILOT written in C language) to control the acquisition procedure during the scan of a sample [275]. It controls the move of the sample stage and starts (or stops) the acquisition computer card. After each point, PILOT reads all the channels of the card and calculate the absolute integral of up to 10 regions of interest (ROI) selected on the spectrum. The results are stored with the co-ordinates of the point in an ASCII file for post treatments (2D or 3D imaging). During the scan, PILOT displays on the computer screen a 2D histogram of the number of counts in the first ROI selected versus the number

Table 6.4

	DCI-D15	ESRF–ID9
X-ray source	bending magnet	undulator
Input energy spectrum	white	11.8 keV (4th harmonic)
Output energy	12.4 keV	11.8 keV
FWHM of the focal spot, μm^2	2×2 (100 μm pinhole) 1×1 (50 μm pinhole)	2.5(V) $\times$ 5(H)
Photon flux, photons s^{-1} μm^{-2} 100 mA^{-1}	6.7×10^4	1.6×10^8

of steps.

The LURE microprobe was tested in different ways and on two different machines: DCI and ESRF. Well known objects were investigated to test both spatial resolution and sensitivity of the microprobe and to learn how to interpret the images.

Photon flux and lens efficiency
(a) *Photon flux.* By measuring the X-ray fluorescence emitted from a thin metallic foil set at the focal spot one can calculate the number of incident photons. At DCI, the total intensity in the spot was measured to be 7×10^5 photons per second for the 100 μm input pinhole for a stored current of 260 mA in the machine. This leads to a flux of 6.7×10^4 photons per second per μm^2 and 100 mA in the band pass of the multilayer for 12.4 keV photons.

At the ESRF the microprobe was installed on beam line 9 (Troika) which is a high-brilliance undulator X-ray source perfectly adapted to our kind of optic.

It measured 2×10^9 photons per second which means a flux of 1.6×10^8 photons per second per μm^2 per 100 mA in the bandpass of the undulator .
(b) *Lens efficiency.* The overall efficiency was measured by comparing the number of photons striking the lens (in the band pass of the multilayer) to that passing in the focal spot for 12.4 keV photons. An efficiency of 16% was obtained that is in agreement with the theoretical prediction for a classical amplitude–phase grating. This indicates the phase character of the obtained diffraction.

Spatial resolution
At this scale spatial resolution starts to be difficult to measure. Photographic plates were used and a knife edge object either in transmission or in fluorescence mode.

(a) *Photographic resolution tests.* High-resolution (0.1 μm VRE type) photographic plates have been used on DCI to test our X-ray microprobe spot configuration. The corresponding microphotographs of the spots with a 50 μm diameter input pinhole are shown in figure 6.6. The size of the observed focal spot was estimated to be about 1 μm in diameter as can be seen from the

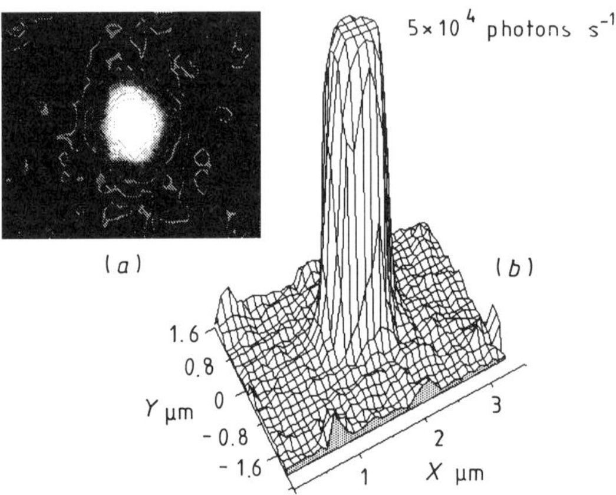

Figure 6.6 (*a*) Microphotographs of the focal spot obtained at DCI with an elliptical Bragg–Fresnel lens and a 50 micron diameter input pinhole. (*b*) Microdensitogram of the photographic plate.

micro-densitogram in figure 6.7.

(b) *Transmission mode resolution test*. The resolution of the X-ray microprobe was further tested in transmission mode. A specially prepared gold mask was used with 5 μm large and 7 μm thick strips, supported on a Si_xC_x membrane. The variation of the absorbed X-ray flux versus position when the strip is moved across the focused beam is shown figure 6.7. A derivation of the transmitted flux is also shown. A resolution of 1 μm was obtained with good contrast. A two-dimensional scanned image of another part of the same gold test object is shown in figure 6.7(*b*). This plot indicates a good signal-to-noise ratio. Data were corrected to take into account the SR beam intensity decay and the position of the sample at 45° to the incident beam.

Even with these corrections figure 6.7(*b*) shows clearly some deviations in the horizontal position of the sample. One of the main differences in instrumentation of such a photon microprobe compared to microprobes using charged particles is that the exciting beam cannot be scanned in position. This constrained us to move the sample itself to perform a scan. Mechanical movements are subject to backlash, and dilatation of the metallic support of the sample can lead to a shift in position, as long as the photon microprobe is set in atmosphere without special temperature stabilization. This may change in the future if surface acoustic waves devices can be used to modulate the X-ray beam position as shown in chapter 4.

(c)*Fluorescence test of a nickel cross*. This object is a cross made of two perpendicular 20 μm diameter nickel wires embedded in a resin. Nickel can be easily excited with a 12.5 keV photon energy. Figure 6.9 shows the Ni profile obtained through a a two-dimensional map of a 150×150 μm^2 area around the

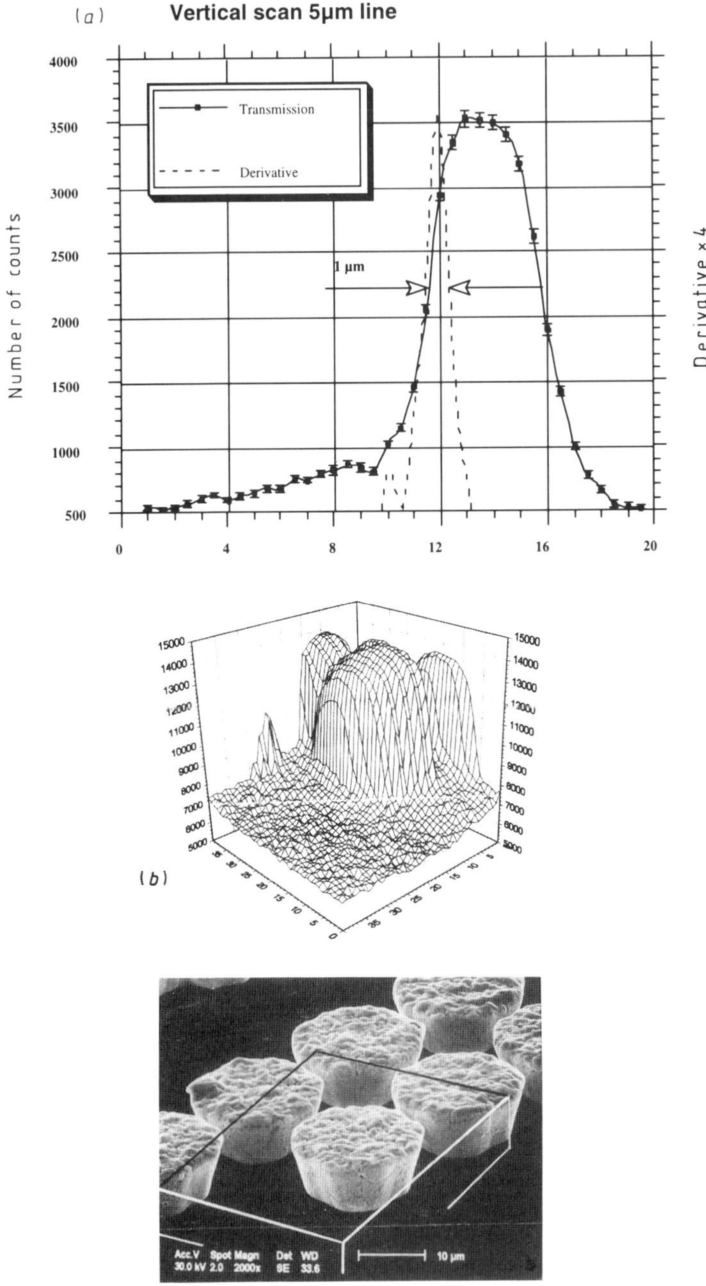

Figure 6.7   (*a*) Vertical scan of a 5 micron strip of the gold mask. (*b*) X-ray microprobe image in transmission of an area of the gold mask. (*c*) Electron microscopic image of the same area.

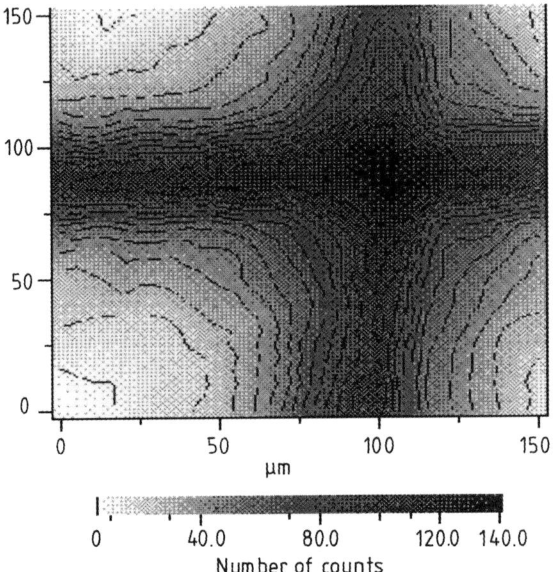

Figure 6.8 Fluorescence of 20 μm diameter Ni wires: 2D map (150×150 μm^2 area around the cross, 15 seconds acquisition per point, 5 μm steps, total map achieved in 4 hours, data corrected to take into account the known exponential decay of the synchrotron beam).

cross. The horizontal scan shows a clear 'shadowing' effect compared to the vertical one. The sample being set at 45° from the beam, when the beam is scanning through the vertical wire there is a continuous variation of two linked parameters which introduces this shadowing: the thickness of the excited part of the sample and auto-absorption of the Ni X-rays.

This result highlights the problem encountered with thick samples. Such a phenomenon is very important when the penetration depth of the exciting X-rays is larger than or close to the thickness of the sample. High energy X-rays are needed to excite fluorescence of heavy elements. But the higher is the energy, the larger is the penetration depth of the exciting beam into the sample. This leads to a reduction in the horizontal spatial resolution of the microprobe with thick samples.

Microprobe calibration
Analysis of calibrated glasses: comparison with proton microprobe (PIXE).
Synthetic glasses of pure oxide were analysed at the ESRF and also at the Pierre S e PIXE microprobe (Laboratoire Pierre S e, LPS, Saclay, France) by M Mosbah *et al* [276]. These glasses contained well known trace concentrations of elements V to Nb. The X-ray fluorescence spectra of one of these glasses observed with both microprobes are shown in figure 6.9.

For a photon microprobe experiment (SRXRF), the excitation energy was set

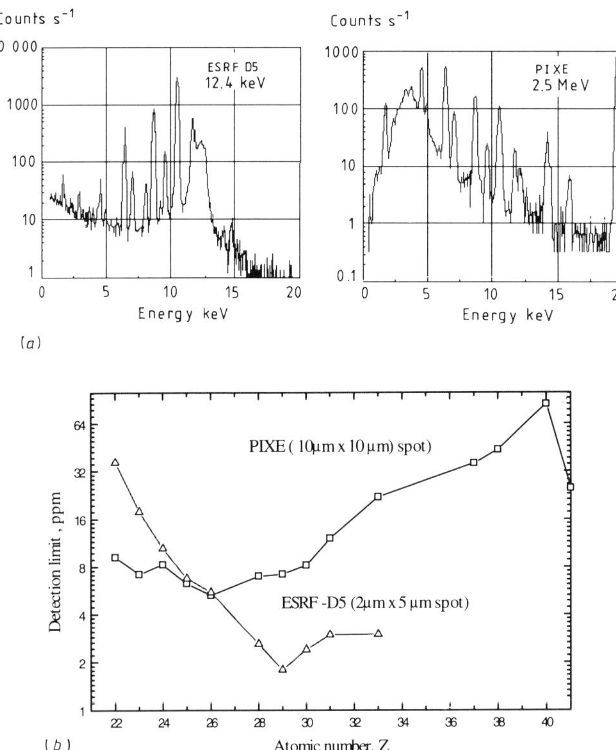

Figure 6.9 (*a*) Spectra of the glass fluorescence. Photon microprobe (left) and proton microprobe (right). (*b*) Practical minimum detection limits, calculated from fluorescense spectrum.

at 11.8 keV but second-order diffracted photons (energy = 23.6 keV) could still excite characteristic lines of elements heavier than Rb with reasonable efficiency. The air path between the sample and the detector (6 cm, solid angle 3.6 mrad) and the rather high excitation energy makes it unfavourable for the observation of light elements. The PIXE spectrum was obtained under 3 μC of 2.5 MeV proton excitation in a vacuum chamber and the solid angle was 13 mrad. To avoid saturation of the detector with the low-energy part of the bremsstrahlung spectrum it is necessary to insert an Al filter between the sample and the detector thus absorbing notably X-rays from the lightest elements.

Up to now, SRXRF and PIXE probes have achieved easily-minimum detection limit (MDL) of 1 to 10 ppm depending on the material and optimal conditions. From the spectra such as the one presented in figure 6.10a one can compute an MDL and compare the sensitivity of both kind of microprobes under usual working conditions (figure 6.10(b).).

This comparison highlights a better sensitivity for elements from V to Fe with

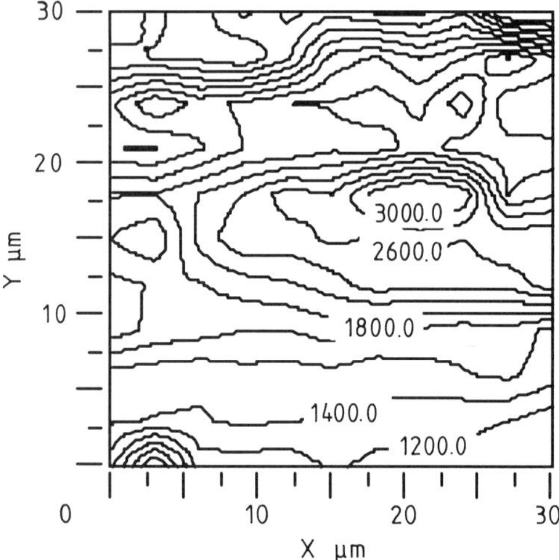

Figure 6.10 Two-dimensional map of the Fe distribution in a micrometeorite sample. Spatial resolution ~ 1 μm.

PIXE and from Ni to As with SRXRF as could be foreseen from the behaviour of the ionization cross section. In fact, as this was the first experiment, we know that MDL reached at the ESRF can be pushed downward by a factor of ten by optimization of the incident energy at the maximum intensity coming from the undulator (a factor of five was measured), better shielding of the detector (perhaps a factor of two) and reduction of the sample-to-detector distance (at least a factor of five). At the same time better MDL for heavy elements could be obtained with micro-PIXE with higher-energy protons but that would probably not change drastically the advantage of photon excitation for this kind of elements.

The microbeam at DCI was used for elemental analysis of small inclusions in micrometeorites collected in Antarctica by melting and filtering large amounts of the blue ice [277]. These micrometeorites are mainly composed of pyroxen and olivine phases (iron and magnesium silicates). In about 60% of these specimens a new phase mainly composed of carbon, oxygen, phosphor and sulphur has been discovered. Due to its major elemental composition this new phase has been called COPS. The formation of this phase is not yet understood and trace element analysis could help to determine the its origin. Due to the small size of these inclusions (10 to 40 μm) only a very sensitive microprobe such as the photon microprobe can be used.

One of the meteorite samples was prepared for a different kind of investigation. It is embedded in epoxy and polished to show a COPS inclusion. Here again the enhanced sensitivity for light elements under charged particle excitation in vacuum was observed. In these siderolite samples, O, Mg, Al and

Si are easily observed with the electron microprobe in a very short exposure time, but the sensitivity drops down for heavy elements. Iron, one of the major constituents, is already best seen under photon excitation (1000 s exposure time). We should also note a better signal-to-background ratio for all elements above Ca with the photon microprobe. Figure 6.10 gives a map of the Fe fluorescence spectral line obtained with X-ray microprobe at DCI for this sample.

6.3   CONCLUSION

We end with a brief summary of promising applications of X-ray diffraction elements in different branches of physics and technology.

Scanning X-ray microscopy.   The first application of X-ray optical elements was in the scanning microscope. Scanning was performed by mechanically displacing the sample in the field of a fixed Fresnel zone plate. Bragg reflection was then used to scan the beam by modulating the scattering power of the Bragg mirror or lens with, for example, an ultrasonic wave.

Local chemical analysis.   Highly localized chemical analysis can be carried out on structures by recording the different secondary phenomena produced by incident X-rays (electrons, secondary X-rays, photoelectrons, absorption). Excitation by X-rays in the course of spectroscopic analysis introduces less damage than exciation by an electron beam.

Image transfer by X-rays.   Image transfer is the traditional problem in optics. In the X-ray range, the transfer of a two-dimensional Fourier image can be performed either by Fourier optics or by a Bragg–Fresnel lenses.

X-ray spectroscopy.   Estimates show that it is possible to construct spectrometers exploiting Bragg diffraction and having a resolving power $\lambda/\Delta\lambda$ of up to 10^6.

X-ray interferometry.   Local X-ray interferometry is possible in addition to traditional applications. It can be used to investigate local defects on the surface of a crystal and in thin surface layers. The sensitivity of the method $\Delta d/d$ is 10^{-7}–10^{-9} for local dimensions of 10 nm.

X-ray lithography.   The advent of devices employing scanning X-ray beams and lenses capable of image transfer in the X-ray range suggests that it will eventually be possible to manufacture microelectronics elements with a resolution of 10–100 nm.

The above list is not exhaustive, but it does present the main applications of X-ray optics to the diagnostics of submicron structures and their fabrication technology in microelectronics.

References

[1]   Templeton D H, Templeton L K, Phillips J C and Hodgson K O 1980 *Acta Crystallogr.* A **36** 436–42

[2]   James R W 1965 *The Optical Principles of the Diffraction of X-rays* (Ithaca, NY: Cornell University Press)

[3]   Henke B L 1981 *Low-Energy X-Ray Diagnostics, Monterey 1981* (AIP Conf. Proc.) pp 146–55

[4]   Cromer D T and Lieberman D 1970 *J. Chem. Phys.* **53** 1881–98

[5]   Henke B L, Lee P, Tanaka I J *et al* 1981 *Low-Energy X-Ray Diagnostics, Monterey 1981* (AIP Conf. Proc.) pp 340–88

[6]   Lukirskii A P 1964 *Doctoral Thesis* Leningrad

[7]   Bonse U, Hartmann-Lotsch I, Lotsch H and Olthoff-Muller K 1982 *Z. Phys.* B **47** 297–9

[8]   Barbee T W, Warburton W K and Underwood J H 1984 *J. Opt. Soc. Am.* **1** 691–8

[9]   Aristov V V, Basov Yu A and Erko A I 1984 *Proc. All-Union Seminar on Microlithography (Chernogolovka)* p 54 (in Russian)

[10] Zimkina T M and Fomichev V A 1971 *Ultrasoft X-ray Spectroscopy* (Leningrad: University Press) p 132 (in Russian)

[11] Sonntag B, Haensel R and Kunz C 1969 *Solid State Commun.* **7** 597–9

[12] Haensel R, Radler K and Sonntag B *ibid.* 1495–7

[13] Jaegle P, Farnoux F C, Dhez P *et al* 1969 *Phys. Rev.* **188** 30–5

[14] Lukirskii A P, Savinov E P, Ershov O A and Shepelev Yu F 1964 *Opt. Spektrosk.* **16** 310–8

[15] Ershov O A, Brytov I A and Lukirskii A P 1967 *ibid.* **22** 127–34

[16] Tanaka, T J Henke B L, Lee P *et al* 1982 *At. Data Nucl. Data Tables* **27**

[17] Blokhin M A and Shveitser A 1982 *X-ray Spectroscopy Handbook* (Moscow: Nauka) (in Russian)

[18] Gluskin E S, Gaponov S V, Dhez P *et al* 1986 *Nucl. Instrum. Methods* A **246** 394–6

[19] Savinov E P, Lyakhovskaya I I, Ershov O A and Kovaleva E A 1969 *Opt. Spektrosk.* **27** 342–7

[20] Filatova O E, Vinogradov A S, Sorokin I A and Zimkina T M 1983 *Fiz. Tverd. Tela* **25** 1280–5

[21] Vinogradov A S, Filatova E O and Zimkina T M *ibid.* 1120–3

[22] Ershov O A and Chernova S I 1969 *Opt. Spektrosk.* **26** 597–600

[23] Tatchyn R and Lindau I 1981 *Low-Energy X-Ray Diagnostics, Monterey 1981* (AIP Conf. Proc.) pp 321–3

[24] Tatchyn R and Lindau I *ibid.* pp 323–6

[25] Bonse U, Hartmann-Lotsch I and Lotsch H *EXAFS and Near Edge Structure* eds A Bianconi, L Incoccia and S Stripcich (Berlin: Springer) pp 362–72

[26] Fontaine A, Warburton W K and Ludwig K F 1985 *Phys. Rev.* B **31** 3599–605

[27] Marmoren R and Andre J M 1983 *Appl. Opt.* **22** 17–9

[28] Hagemann H J, Gudat W and Kunz C 1974 *Desy Lab. Rep.* N SR74/7 (Hamburg) 48

[29] Andre J M, Maguet A and Barchewitz R 1982 *Phys. Rev.* B **25** 5671–9

[30] Labov S, Bowyer S and Steele G 1985 *Appl. Opt.* **24** 576–8

[31] Khare B N, Sagen C, Arakawa E T *et al* 1984 *Icarus* **60** 127–37

[32] Jentzsche F 1929 *Phys. Z.* **30** 268–73

[33] Michette A G 1986 *Optical Systems for Soft X-rays* (New York: Plenum) p 310

[34] Ehrenberg W 1947 *Nature* **160** 330–1

[35] Kirkpatrick P and Baez A V 1948 *J. Opt. Soc. Am.* **38** 766–74

[36] Price R H 1981 *Low-Energy X-ray Diagnostics, Monterey 1981* (AIP Conf. Proc.) p 189

[37] McGee J F 1957 *X-ray Microscopy and Microradiography* (New York) p 164

[38] McGee J F, Hesser D R and Miton J W 1969 *X-ray Reflection Optics (Recent developments)* (Berlin: Springer) p 251

[39] Seward F D and Pelmier T M 1975 *Rev. Sci. Instrum.* **46** 204–6

[40] Seward F D, Dent J, Boyle M *et al* 1976 *ibid.* **47** 464–70

[41] Wolter H 1952 *Ann. Phys., Lpz* **10** 94–114

[42] Franks A 1977 *Sci. Progr.* **64** 371–422

[43] Arkadev V A, Kolomiitsev A I, Kumakhov M A *et al* 1987 *Poverkhnost'. Fiz. Khim. Mekh.* **2** 44–7

[44] Aleksandrov Yu A, Valiev K A, Velikov A V *et al* 1987 *Pis'ma Zh. Eksp. Teor. Fiz.* **13** 257–60

[45] Myers O E 1951 *Am. J. Phys.* **19** 359–65

[46] Baez A V 1952 *J. Opt. Soc. Am.* **42** 756–62

[47] Baez A V 1961 *ibid.* **51** 405–9

[48] Born M and Wolf E 1970 *Principles of Optics* (Oxford: Pergamon) p 719

[49] Horman H and Chau H M 1967 *Appl. Opt.* **6** 317

[50] Keating R N, Mueller R K and Sawatari J 1972 *J. Opt. Soc. Am.* **62** 945–50

[51] Rarback H, Kenney J M, Kirtz J *et al* 1984 *X-ray Microscopy* ed G Schmal and D Rudolph (Berlin: Springer)

[52] Waldman G S 1966 *J. Opt. Soc. Am.* **56**

[53] Stiglian D J, Mittra R and Semonin R D 1967 *ibid.* **57** 610–3

[54] Korn G A and Korn T M 1961 *Handbook of Mathematics* (New York: McGraw-Hill)

[55] Stroke G W 1965 *Phys. Lett.* **16** 272–3

[56] Samson J A R 1967 *Techniques of Vacuum Ultraviolet Spectroscopy* (Berlin: Springer) p 218

[57] Thieme J 1984 *X-ray Microscopy* eds G Schmal and D Rudolph (Berlin: Springer) p 121

[58] Niemann B 1986 Soft X-ray Optics and Technology *SPIE Conf. Proc.* **733** pp 422–427

[59] Young M 1972 *J. Opt. Soc. Am.* **62** 972–6

[60] Guttmann P 1984 *X-ray Microscopy* eds G Schmal and D Rudolph (Berlin: Springer)

[61] Simpson M J and Michette A G 1984 *Opt. Acta* **31** 403–13

[62] Lord Rayleigh 1888 *Encyclopaedia Britannica* 9th edn vol 24 p 429

[63] Wood R W 1988 *Phil. Mag.* **45** 511

[64] Miyamoto K 1961 *J. Opt. Soc. Am.* **51** 17–21

[65] Kirz J 1974 *ibid.* **64** 301–9

[66] Ceglio N M, Hawryluk and A M Schattenburg M 1983 *J. Vac. Sci. Technol.* B **1**

        1285–9

[67]   Davydov A V, Panchenko L A, Erko A I 1987 *Pis'ma Zh. Tekh. Fiz.* **13** 1017–20
[68]   Gornakova N V, Bashkina G A, Davydov A V and Erko A I 1986 *Proc. All-Union Conf. on Physical Methods for Studying Surfaces and the Diagnostics of Materials and Computer Components* (Kishinev) p 35 (in Russian)
[69]   Horman M H 1967 *Appl. Opt.* **6** 2011
[70]   Hilkenbach R and Thieme J 1986 *Soft X-ray Optics and Technology* (SPIE Conf. Proc.) **733** 464–70
[71]   Burge R E, Michette A G, Browne M T *et al ibid.* 460
[72]   Schmahl D, Rudolph D, Guttmann P and Christ O 1984 *X-Ray Microscopy* (Berlin: Springer) p 63
[73]   Lesem L B, Hirsch P M and Jordan J A Jr 1969 *IBM J. Res. Devel.* **13** 150–8
[74]   Tatchyn R, Csonka P L and Lindau I 1982 *J. Opt. Soc. Am.* **72** 1630–34
[75]   Tatsyn R, Csonka P L and Lindau I 1983 *IEEE J. Quantum Electron.* **QE-19** 1821–3
[76]   Tatchyn R 1984 *X-Ray Microscopy* ed G Schmal and D Rudolph (Berlin: Springer) pp 40–50
[77]   Velikhov E P and Prokhorov A P 1984 *Computer Optics* (Moscow: MTsNTI) **1** (in Russian)
[78]   Koronkevich V P, Lenkova G A, Mikhailisova I A *et al* 1985 *Avtometriya* **1** 4–25
[79]   Bashkina G A, Babin S V, Davydov A V *et al* 1988 *Mikroelektronika* **17** 184–6
[80]   Kyurgai H and Urisi T 1986 *Nucl. Instrum. Methods Phys. Res.* A **245** 199
[81]   Aristov V V Erko and A I Panchenko *et al X-ray Microscopy II* eds D Sayre, M Howells, M Kirz and H Rarback (Berlin: Springer) p 138
[82]   Collier R J, Burckhardt C B and Lin L H 1971 *Optical Holography* (Academic)
[83]   Aristov V V and Ivanova G A 1979 *J. Appl. Crystallogr.* **12** 19–24
[84]   Aristov V V, Polovinkina V I, Shmyt'ko I M and Shulakov E V 1978 *Pis'ma Zh. Tekh. Fiz.* **28** 6–9
[85]   Kikuta S Aoki S Kosaki S and Kohra K 1972 *Opt. Commun.* **5** 86–9
[86]   Saecocio E J 1967 *J. Opt. Soc. Am.* **57** 966–70
[87]   Aoki S and Kikuta S 1974 *Japan. J. Appl. Phys.* **13** 1385–92
[88]   Aristov V V, Bashkina G A and Erko A I 1980 *Phys. Status. Solidi* a **59** 663–71
[89]   Meier R W 1965 *J. Opt. Soc. Am.* **55** 987–92
[90]   Feder R, Spiller E, Topalian J *et al* 1977 *Science* **197** 259–60
[91]   Bjorkland G C, Harris S E and Joung J E 1974 *Appl. Phys. Lett.* **25** 451–2
[92]   Aristov V V, Bashkina G A and Erko A I 1980 *Opt. Commun.* **34** 332–6
[93]   Aristov V V and Erko A I 1981 *Thirteenth All-Union School on Holography* (Leningrad) p 191 (in Russian)
[94]   Fay B 1981 *J. Vac. Sci. Technol.* **19** 1194–9
[95]   Attwood D, Halbach K and Kim K-J 1985 *Science* **228** 1265–72
[96]   Kondratenko A M and Skrinsky A N 1975 *The Use of Radiation from Storage Rings in X-ray Holography of Micro-objects* (Novosibirsk) Preprint INP-N 102 (in Russian)
[97]   Howells M R 1987 *X-Ray Microscopy* eds G Schmal and D Rudolph (Berlin: Springer)
[98]   Gluskin E S *et al ibid.*
[99]   Economou N P and Flanders D C 1981 *J. Vac. Sci. Technol.* **19**
[100]   Cowley J M 1977 *Diffraction Physics* (Amsterdam: North-Holland)
[101]   Denisyuk Yu N, Romishvili N M and Chavchanidze V V 1971 *Opt. Spektrosk.* **30** 1130–4
[102]   Smirnov A P 1979 *ibid.* **46** 574–8

[103] Lokshin G P and Lepekhin V D 1983 *ibid.* **54** 705–10
[104] Semenov A T 1979 *Kvant. Elektron.* **6** 1804–7
[105] Flanders D C Hawryluk and Smith H I 1979 *J. Vac. Sci. Technol.* **16** 1649–952
[106] Jahns Y and Lohmann A W 1979 *Opt. Commun.* **28** 263–7
[107] Gori F 1979 *ibid.* **31** 4–8
[108] Sudol R and Thompson B Y *ibid.* 105–10
[109] Brenner K H, Lohmann A W and Ojeda-Castaneda J 1983 *ibid.* **46** 14–7
[110] Bell R J 1972 *Introductory Fourier Transform Spectroscopy* (New York: Academic)
[111] Aristov V V, Bashkina G A, Dorozhkina L V *et al* 1983 *Poverkhnost'. Fiz. Khim. Mekh.* 113–8
[112] Aristov V V, Erko A I and Martynov V V 1988 *Opt. Spektrosk.* **64** 630–7
[113] Aristov V V, Erko A I and Kopetskii, 1982 *Proc. Int. Conf. on Microlithography: ME-82, Grenoble* ch 5 pp 137–41
[114] Aristov V V, Erko A I and Martynov V V 1984 *Proc. All-Union Seminar on Microlithography, Chernogolovka* p 50 (in Russian)
[115] Aristov V V, Aoki S, Erko A I *et al* 1985 *Opt. Commun.* **56** 223–55
[116] Aristov V V, Erko A I and Kudriashov V A 1985 *Microelectron Eng.* **3** 589–95
[117] 1984 *Photon Factory Activity Report for 1982/83* Nat. Lab. High-Energy Phys. KEK, p IY-50
[118] Aristov V V, Erko A I and Martynov V V 1985 *3rd All-Union Conf. on Coherent Interaction of Radiation with Matter, Uzhgorod* (in Russian)
[119] Aristov V V, Erko A I and Martynov V V 1986 *Short Wavelength Coherent Radiation: Generation and Application* AIP Conf. Proc. eds D Attwood and J Bocor p 71
[120] Aristov V V, Erko A I and Martinov V V 1985 *Opt. Commun.* **53** 159–63
[121] Brynagahl O 1973 *J. Opt. Soc. Am.* **63** 416–9
[122] Lokshin G R, Belonuchkin V E and Kozel S M 1985 *Holographic Methods in Science and Technology* (Leningrad) p 47 (in Russian)
[123] von Deudner V 1930 *Ann. Phys.* **5** 261–80
[124] DuMond J and Yowtz J P 1940 *J. Appl. Phys.* **11** 357–65
[125] Dinklage J 1967 *ibid.* **34** 2633–5
[126] Dinklage J 1967 *ibid.* **38** 3781–5
[127] Spiller E 1972 *J. Appl. Phys. Lett.* **20** 365–7
[128] Spiller E 1976 *Appl. Opt.* **15** 2333–8
[129] Spiller E 1981 *Low-Energy X-ray Diagnostics, Monterey 1981* (AIP Conf. Proc.) p 124
[130] Barbee T W *ibid.* 131–45
[131] Vinogradov A V and Zel'dovich B Ya 1977 *Opt. Spektrosk.* **42** 709–14
[132] Vinogradov A V and Zel'dovich B Ya 1977 *Appl. Opt.* **16** 89–93
[133] Saxena A M and Schoenborn B P 1977 *Acta Crystallogr.* A **33** 805–13
[134] Underwood J H and Barbee T W 1981 *Low-Energy X-Ray Diagnostics, Monterey 1981* AIP Conf. Proc.) 170–8
[135] Gaponov S V, Genkin V M, Salashchenko N M and Fraerman A A 1985 *Pis'ma Zh. Eksp. Teor. Fiz.* **41** 53–5
[136] Kogelnik H 1969 *Bell. Syst. Technol. J.* **48** 2909–47
[137] Beckmann P 1963 *The Scattering of Electromagnetic Waves by Rough Surfaces* Part 1 (Oxford: Pergamon)
[138] Akhsakhalyan A D, Gaponov S V, Gusev S A *et al* 1984 *Zh. Tekh. Fiz.* **54** 755–62
[139] Ziegler E, Lepetre Y, Schuller I K and Spiller E 1986 *Appl. Phys. Lett.* **48**
[140] Bartlett R Y, Kania D R, Trela W J *et al* 1984 *Opt. Eng.* **23** 667
[141] Ziegler E, Lepetre Y and Fenske G 1986 *Multilayer Structures and Laboratory*

156   REFERENCES

         X-ray Laser Research (SPIE Conf. Proc.) **688** 193–7
[142] Spiller E 1988 *Rev. Phys. Appl.* **23** 1687
[143] Pianetta P, Barbee T W and Redaelli R 1986 *Nucl. Instrum. Meth. Phys.* A **246** 352–5
[144] Chauvineau J P, Cornon Y, Naccache P *et al* 1984 *Nucl. Instrum. Meth.* **208** 227–31
[145] Gaponov S V, Garin F V, Gusev S L *et al* 1984 *Nucl. Instrum. Meth.* **208** 227–31
[146] Gaponov S V, Gusev S L, Platonov Yu Ya and Salashchenko N N 1984 *Zh. Tekh. Fiz.* **54** 747–54
[147] Dhez P 1986 Soft X-ray Optics and Technology *SPIE Conf. Proc.* **733** 308–15
[148] Dhez P 1985 *J. Microsc.* **138** 267–77
[149] Dhez P 1983 *Adv. Space Res.* **2** 199–206
[150] Chauvineau J P, Marioge J P, Bridou F *et al* 1986 Soft X-ray Optics and Technology *SPIE Conf. Proc.* **733** 301–6
[151] Thompson A C, Wu Y and Underwood J H 1987 *Nucl. Instrum. Methods Phys. Res.* A **255** 603–5
[152] Marshall G F 1986 *Opt. Eng.* **25** 922
[153] Aristov V, Erko A I and Martynov V V 1988 *Rev. Phys. Appl.* **23** 1623–30
[154] Barbee T W 1984 *X-ray Microscopy and Structure* eds G Schmal and D Rudolph (Berlin: Springer)
[155] Ceglio N M, Stearns D G and Hawryluk A M 1985 Applications of Thin-Film Multilayered Structures to Figured X-ray Optics *SPIE Conf. Proc.* **563** p 360
[156] Denisyuk Yu N 1963 *Opt. Spektrosk.* **15** 522–30
[157] Aristov V V and Shekhtman V Sh 1971 *Usp. Fiz. Nauk* **104** 51–7
[158] Aristov V V, Erko A I, Snigirev A A and Niculin A N 1986 *Opt. Commun.* **58** 300–2
[159] Aristov V V, Basov Yu A and Snigirev A A 1987 *Pis'ma Zh. Tekh. Fiz.* **13** 114–8
[160] Aristov V V, Snigirev A A, Basov Yu A and Niculin A Yu 1986 Short-Wavelength Coherent Radiation: Generation and Application *AIP Conf. Proc.* **147** 253–8
[161] Aristov V V, Gaponov S V, Genkin V M *et al* 1986 *Pis'ma Zh. Eksp. Teor. Fiz.* **44** 207–9
[162] Aristov V V, Gaponov S V, Erko A I and Salashchenko N N 1986 *Opt. News* **12** 128
[163] Erko A I 1988 *Proc. Second All-Union Seminar on Microlithography, Chernogolovka* (in Russian)
[164] Agaronov Yu A, Gorbatov Yu B, Erko A I *et al* 1988 *ibid.*
[165] Erko A I 1989 *Rev. Sci. Instrum.* **60** 2502
[166] Aristov V V, Erko A I, Firsov A A *et al* 1989 *Abstracts of 2nd Eur. Conf. on Advances in X-Ray Synchrotron Radiation Research, Rome* pp 6–35
[167] Entin I R 1985 *Phys. Status Solidi* b **132** 355–64
[168] Mkrtchyan A R, Navasaryan M A, Gabrielyan R G *et al* 1983 *Pis'ma Fiz. Tekh. Fiz.* **9** 1181–3
[169] Gabrielyan R G and Aslanyan H A 1984 *Phys. Status Solidi* b **82** 35–45
[170] Cerva H and Graeff W 1984 *Phys. Status Solidi* a **123** 1197–9
[171] Cerva H and Graeff W 1986 *ibid.* **93** 1129–32
[172] Kikuta S, Takahoshi T and Nakatani S 1984 *Japan. J. Appl. Phys.* **23** 193–6
[173] Aristov V V, Vereshchagin G V, Erko A I *et al* 1987 *Pis'ma Zh. Tekh. Fiz.* **13** 1288–91
[174] Vereschagin G V, Erko A I, Roschupkin D V *et al* 1989 *Nucl. Instrum. Methods Phys. Res.* **282** 634–7
[175] Erko A I 1987 *Materials for the Computer Technology* (Moscow) p 67 (in Russian)

[176]   Schmahl G and Rudolph D 1970 *Optik* **30** 606–9
[177]   Schmahl G, Rudolph D, Guttmann P and Christ O 1987 *X-ray Microscopy* eds
        G Schmal and D Rudolph (Berlin: Springer)
[178]   Gabor D 1948 *Nature* **161** 777-8
[179]   Rudolph D, Schmahl G and Niemann B *et al* 1987 *Phys. Scr.* T **17** 201–3
[180]   Rudolph D 1986 Soft X-ray Optics and Technology *SPIE Conf. Proc.* **733** 294–
        300
[181]   Rudolph D, Niemann B and Schmahl G 1982 Status of the Sputtered Sliced Zone
        Plate for X-ray Microscopy *SPIE Conf. Proc.* **316** 103–5
[182]   Saitoh K, Inagawa K, Kohra K *et al* 1988 *SPIE Conf. Proc.* **316** pp 103–5
[183]   Brown W L, Venkatejan T and Nager A 1981 *Nucl. Instrum. Meth.* **191** 157–68
[184]   Lewis G, Mioduszewski J, Weiner D and Siegel B 1988 *J. Vac. Sci. Technol.* B
        6 239–40
[185]   Horiuchi K, Itakura T and Ishihara H *ibid.* 241–4
[186]   Sudraud P, Ben Assayag G and Bon M *ibid.* 234–8
[187]   Morimoto H, Sasaki Y, Saitoh K *et al* 1986 *Microelectron. Eng.* **4** 163–79
[188]   Ringer M, Hidber H R, Schlogl R *et al* 1985 *Appl. Phys. Lett.* **46** 832–4
[189]   Rothschild M and Ehrlich J 1988 *J. Vac. Sci. Technol.* B **6** 1–17
[190]   Spiller E and R Feder 1978 *Sci. Am.* **239** 60–8
[191]   Valiev K A and Rakov A V 1984 *Physical Principles of Submicron Lithography
        in Microelectronics* (in Russian) (Moscow: Radio i Svyaz') p 350
[192]   Valiev K A 1986 *Microelektronics: Advances and Perspectives* (Moscow: Nauka)
        (in Russian)
[193]   Brodie I and Muray J J (1982) 1985 *The Physics of Microfabrication* (New York:
        Plenum)
[194]   Aristov V V, Kopetskii Cn V, Kokhanchik G I and Kudryashov V A 1983
        Poverkhnost'. Fiz. Khim. Mekh. **11** 5–15
[195]   Smith H I and Bernacki S E 1975 *J. Vac. Sci. Technol.* **12** 1321–3
[196]   Vladimirsky Yu 1988 *J. Vac. Sci. Technol.* B **6** 183–5
[197]   Urisu T and Kyuragi H 1987 *ibid.* **5** 1436–40
[198]   Mott N and Massey H S W 1967 *The Theory of Atomic Collisions* (Oxford:
        Claredon)
[199]   Fontaine G 1969 *Microanalysis and Scanning Electron Microscopy* (Moscow:
        Mir) (in Russian translation)
[200]   Chang T H P, Kern D P, Kratschmer E *et al* 1988 *IBM J. Res. Dev.* **32** 1436–40
[201]   Aritome H, Nagata K and Namba S 1985 *Microelectron. Eng.* **3** 459–66
[202]   Aritome H and Namba S 1986 *SPIE Conf. Proc.* **733** 440–8
[203]   Aritome H, Matsui S, Moriwaki K *et al* 1983 *Nucl. Instrum. Methods* **208** 233
[204]   Burge R E, Michette A G, Broune M I *et al* 1986 *SPIE Conf. Proc.* **733** 460–463
[205]   Kratchmer E, Beneking H, Erko A I and Petrashov V T 1984 *Appl. Phys. Lett.*
        44 1011–3
[206]   Bogli V, Unger P and Beneking H 1986 *SPIE Conf. Proc.* **733** 449–58
[207]   Vladimirsky Y, Kern D and Chang T H P *et al* 1988 *J. Vac. Sci. Technol.* B **6**
        311–5
[208]   Muray A, Issacson M, Adesida I and Whitehead B 1983 *ibid.* **1** 1091–5
[209]   Koops H W P and Grob J 1984 *X-Ray Microscopy* eds G Schmal and D Rudolph
        (Berlin: Springer)
[210]   Browne M T, Charalambous P, Burge R E *et al* 1984 *J. Phys. C: Solid State
        Phys.* **245** 89–95
[211]   Koops H W P, Weiel R, Kern D P and Baum T H 1988 *J. Vac. Sci. Technol.* B
        6 477–81
[212]   Babin S V, Erko E I, Dubovitskaya E P and Smirnov V A 1986 *Proc. All-

Union Conf. on Physical Methods for Studying Surfaces and the Diagnostics of Materials and Computer Components, Kishinev p 68 (in Russian)

[213]   Valiev K A, Kirillov A N, Kovtun B N *et al* 1985 *Preprint IOFAN* (Moscow) 34pp

[214]   Valiev K A, Velikov A V, Kirillov A N *et al* 1986 *Mikroelektronika* **15** 61–5

[215]   Valiev K A, Kirillov A N, Kovtun B N *et al* 1987 *ibid.* **16** 122

[216]   Derkach V P, Merzhvinskii A A and Starikova L V 1985 *ibid.* **14** 467–77

[217]   Kuser D F and Ting C H 1979 *J. Vac. Sci. Technol.* **16** 1726–33

[218]   Aizaki N 1979 *ibid.* **16** 1759–63

[219]   Phang J C H and Ahmed H *ibid.* 1754

[220]   Kazo I F, Kosunskii V M and Kurbatskaya N P 1984 *Physics and Technology of Cybernetics* (Kiev: IK AN SSSR) p 15 (in Russian)

[221]   Parikh M 1980 *IBM J. Res. Develop.* **24** 438–51

[222]   Valiev K A, Kovtun B N, Kudrya V P and Makhvikadze T M 1987 *Physical and Technological Problems in Cybernetics* ed D Yu Zanyavichus (Vilnius: Lithuanian Academy of Sciences) p 3 (in Russian)

[223]   Aristov V V, Erko A I, Babin S V and Dorozhkina L V 1983 *Proc. Int. Conf. on Microlithography-ME 83, Cambridge* pp 56–63

[224]   Aristov V V, Erko A I and Babin S 1985 *Proc. Int. Conf. on Electron Beam Technology, Varna* (in Russian) p 501

[225]   Babin S V, Erko A I, Roshchupkin D V and Svintsov A A 1988 *Poverkhnost'. Fiz. Khim. Mekh.* **3** 144

[226]   Zaitsev S I and Svintsov A A 1986 *ibid.* **4** 27

[227]   Zaitsev and S I Svintsov A A 1987 *ibid.* **1** 47

[228]   Aristov V V, Babin S V, Davydov A V *et al* 1987 *Microelectron. Eng.* **6** 129–34

[229]   Babin S V, Erko A I and Svintsov A A 1988 *Proc. Int. Conf. on Electron Beam Technology EBT-88, Varna* (in Russian) p 379

[230]   Grun A E 1957 *Z. Naturf. A* **12** 89–95

[231]   Babin S V, Erko A I, Davydov A V and Svintsov A A 1988 *Poverkhnost'. Fiz. Khim. Mekh.* **4** 79–82

[232]   Dremova N N Erko A I and Roshchukin D V 1987 *ibid.* **7** 125–30

[233]   Aristov V V, Agafonova V A, Erko A I and Roshchukin D V 1985 *Proc. Int. Conf. on Electron Beam Technology, Varna* p 468

[234]   Dutley E H 1970 *Semiconductors and Semimetals* (New York: Academic)

[235]   Babin S V, Davydov A V and Erko A I 1987 *Proc. Int. Conf. Akustoelektronika-87, Varna* (in Russian) p 121

[236]   Babin S V Davydov A V and Erko A I 1987 *Prib. Tekh. Eksp.* **2** 191–85

[237]   Davydov A V and Erko A I 1988 *Proc. Int. Conf. on Electron Beam Technology, Varna* **3** 386–8 (in Russian)

[238]   Babin S V and Erko A I 1989 *Nucl. Instrum. Methods Phys. Res.* A **282** 529–31

[239]   Shiono T, Setsune K, Yamazaki O and Wasa K 1987 *Appl. Opt.* **26** 587–91

[240]   Carlson T A 1975 *Photoelectron and Auger Spectroscopy* (New York: Plenum)

[241]   Mundschau M 1991 *Synchrotron Rad. News* **4** 29–33

[242]   Cazaux J 1984 *Ultramicroscopy* **12** 321–32

[243]   Masujima T, Shiwaku H, Yoshida H, Sano T, Amemiya Y, Kawata H, Kataoka M, Hoshi M, Nagoshi C, Uehara S, Imai H and Ando M 1989 *Rev. Sci. Instrum.* **60** 2468–71

[244]   Rarback H, Shu D, Su Cheng Feng, Ade H, Jacobsen C, Kirz J, McNulty I, Vladimirski Y, Kern D and Chang P 1988 *X-ray Microscopy II* eds D Sayre, M Howells, J Kirz and H Rarback *Springer Series in Optical Sciences* **56** 194–200

[245]   Rudolph D, Schmal G, Niemann B, Meyer-Illse W, Guttmann P and Nyakatura

G 1987 *Phys. Scr.* **17** 201–3

[246] Kagoshima Y, Aoki S, Kakuchi M, Sekimoto M, Maezawa H, Hyodo K and Ando M 1989 *Rev. Sci. Instrum.* **60** 2448–52

[247] Kirz J and Rarback H, 1989 *Rev. Sci. Instrum.* **56** 1–13

[248] Browne M T, Burge R E, Charalambous, Duke P J, Freake A, McDowell A, Michette A, Nave G, Rosser R, Simpson M J and Smith W 1984 *J. Physique Coll. C2 suppl 2* **45** 87–8

[249] Michette A G 1984 *J. Photogr. Sci.* **32** 45–8

[250] Rudolph D, Neimann B, Schmahl G and Thieme J 1988 *X-ray Microscopy II* eds D Sayre, M Howells, J Kirz and H Rarback *Springer Series in Optical Sciences* **56** (New York: Springer) 216–9

[251] Trail J A, Byer R L and Kortright J B *ibid.* 310–315

[252] Nyakatura G, Meyer-Illse W, Guttmann P, Niemann B, Rudolph D, Schmahl G, Sarafis V, Hertel N, Uggerhøj E, Skriver E, Nørgaard J O R and Maunsbach A B *ibid.* 365–71

[253] Cazaux J 1975 *Rev. Appl. Phys.* **10** 263

[254] Hovland C 1977 *Appl. Phys. Lett.* **30** 264

[255] Jates K and West R 1983 *Surf. Interface Anal.* **5** 217

[256] Chancy R L 1987 *ibid.* **10** 36–47

[257] Pollack F, Lowventhal S, Palippou D and Fournet P 1988 *X-ray Microscopy II* eds D Sayre, M Howells, J Kirz and H Rarback *Springer Series in Optical Sciences* **56** 220–7

[258] King P L, Browning R, Pianetta P, Lindau I, Keenlyside M and Knapp G 1989 *Rev. Sci. Instrum.* **60** 1686–9

[259] Jacobsen C, Kenney J M, Kirz J, Rosser R J, Cinotti F, Rarback H and Pine Y J 1987 *Phys. Med. Biol.* **32** 431–7

[260] Petersen H, Braun W, Koch E-E and Rudolph D 1986 *SPIE Proc.* **733** 428–431

[261] Sparks C J and Ice G E, 1989 *Mat. Res. Soc. Symp. Proc.* **143** 223–33

[262] Horowitz P and Howell J A 1972 *Science* **178** 608

[263] Jones K W, Kwiatek W M, Gordon B M, Hanson A L, Pounds J G, Rivers M L, Sutton S R, Thompson A C, Underwood J H, Giauque R D and Wu Y 1988 *Advances in X-ray Analysis* eds C S Barnell (New York: Plenum)

[264] Wu Y, Thompson A C, Underwood J H, Giauque R D, Chapman K, Rivers M L and Jones K W 1990 *Nucl. Instrum. Methods* A **291** 146

[265] von Langevelde F, Bowen D K, Tros G H J, Vis R D, Huizing A and de Boer D K G 1990 *Nucl. Instrum. Methods* A **292** 719–27

[266] Erko A, Khzmalian E, Panchenko L, Redkin S, Zinenko V, Chevallier P, Dhez P, Khan-Malek C, Freund A and Vidal B 1990 *Proc. 4th Conf. on X-ray Microscopy (London)*

[267] Erko A, Agafonov Yu, Panchenko L A, Yuakshin A, Chevallier P, Dhez P and Legrand F 1994 *Opt. Commun.* **106** 146–50

[268] Snigirev A *et al* 1995 *Rev. Sci. Instrum.* **66** 1461

[269] Erko A, Chevallier P, Dhez P, Legrand F and Vidal B 1992 *Inst. Phys. Conf. Ser. 130 Proc. XIII Int. Conf. on X-Ray Optics and Microanalysis* (Bristol: Institute of Physics Publishing) p 617

[270] Vidal B and Vincent P 1984 *Appl. Opt.* **23** 1794–801

[271] Vidal B and Dhez P 1988 *SPIE* **688** 110

[272] Vidal B, Jiang Z and Samuel F 1992 *SPIE* **1738** 31

[273] Vincent P and Vidal B 1994 *Proc. 4th Int. Conf. on X-ray Microscopy* Institute of Microelectronics Technology RAS (Chernogolovka) pp 87–102

[274] Mirone A, Idir M, Dhez P, Soullie G, Erko A 1994 *Opt. Commun.* **111** 191–8

[275] Legrand F 1995 *These Docteur ét Science* Université de Paris-Sud, Centre

d'Orsay, France
[276] Mosbah M, Clocchiatti R, Michaud V, Piccot D, Chevallier P, Legrand F, Als Nilsen G and Grübel G 1994 *4th Int. Conf. on Nuclear Microprobe Technology and application (Shangai)*
[277] Legrand F, Erko A, Dhez P, Chevallier P and Engrand C 1994 *Proc. 4th Int. Conf. on X-ray Microscopy* Institute of Microelectronics Technology RAS, Chernogolovka, pp 136–41

Index